T0341308

Intensifying Activated Sludge Using Media-Supported Biofilms

Intensifying Activated Sludge Using Media-Supported Biofilms

Dwight Houweling
Glen T. Daigger

CRC Press
Taylor & Francis Group
Boca Raton London New York

CRC Press is an imprint of the
Taylor & Francis Group, an **informa** business

Cover image: Installation of a ZeeLung cassette into an activated sludge bioreactor. Artistic rendering courtesy of SUEZ Water Technologies & Solutions.

CRC Press
Taylor & Francis Group
6000 Broken Sound Parkway NW, Suite 300
Boca Raton, FL 33487-2742

Printed on acid-free paper

International Standard Book Number-13: 978-0-367-20227-9 (Hardback)

Visit the Taylor & Francis Web site at
http://www.taylorandfrancis.com

and the CRC Press Web site at
http://www.crcpress.com

Dedication

The authors would like to dedicate this work to their families and colleagues who have supported them and collaborated on realizing the disruptive potential of hybrid biofilm/suspended growth processes. We work in multi-disciplinary teams where each person brings a different piece to the puzzle. And it's only through teamwork that we can do anything meaningful or exciting. So thank you to all of you who have contributed, both personally and professionally, to making this book possible. This work is also dedicated to the innovators who see how much more there is to do and who are not constrained by conventional practice. Conventional practice has much to teach us about how technology functions, and needs to function, in the real world. But simply continuing with solutions of the past does not suffice. Successful innovators build on our evolving scientific and engineering knowledge while striving to employ this knowledge on higher performing approaches that, in time, redefine conventional practice and advance how our profession serves society and the planet.

Contents

Preface

A long history of integrating suspended growth and biofilm processes exists in the wastewater profession. The Hayes process provides an early example: wooden planks were added to conventional activated sludge tanks to provide biofilm surface area to increase the biomass contained in the system and hence its treatment capacity. I began working with such systems in the late 1970s and early 1980s when coupling the traditional trickling filter process with suspended growth systems was quite popular due to their simplicity and energy-efficiency. Suspended growth processes have historically demonstrated the capability to produce a high-quality effluent, low in suspended and dispersed solids, while biofilm processes offer more compact treatment because of their generally higher volumetric treatment efficiency. Many coupled, or hybrid, processes have been developed and demonstrated, with the objective of retaining the advantages of each process component while minimizing their disadvantages.

Today we understand that suspended growth processes provide the capability to effectively metabolize a wide range of constituents, including particulate, colloidal, and dissolved organic matter. The selective pressure imposed by the settling and recycling of biomass, together with controlled wasting, result in the development of a well-flocculated solids fraction that produces an effluent low in suspended solids. In contrast, biofilm processes efficiently metabolize dissolved constituents, including dissolved organic matter, which can result in the growth of excessive quantities of filamentous organisms when proper conditions are not provided in a suspended growth bioreactor. Experience with the integrated fixed film activated sludge (IFAS) process has taught us that hybrid systems can provide an increased diversity of environmental niches to selectively retain necessary biological populations under operating conditions not previously possible. For example, nitrifiers will selectively populate the biofilm of IFAS processes when the suspended growth aerobic solids residence time (SRT) is insufficient to allow these slower growth organisms to reproduce on their own in the suspended growth bioreactor. The potential for aerobic and anoxic environments to be created in the biofilm attached to the media in IFAS systems also creates additional process capabilities, analogous to what can also occur in suspended growth process flocs (simultaneous biological nutrient removal).

Incorporating membrane aerated biofilm reactors (MABRs) into suspended growth processes represents an extension of this long history with coupled biofilm and suspended growth processes while providing additional advantages. The greatly increased oxygen transfer efficiency with MABR processes represents an economically important and obvious advantage. The capability to create a further range of environmental niches, for example an aerobic environment on the inside of the biofilm, represents a further advantage. In this work we explore some of these advantages, based on our current understanding of biofilm and suspended growth processes coupled with evolving practical experience. It is hoped that the insights we are able to provide here provide the intellectual "seed" for innovators to greatly expand how process options, such as the coupled MABR/activated sludge process, can be further improved, thereby serving people and the planet.

Glen T. Daigger, Ann Arbor, Michigan, 3 March 2019

I certainly haven't been working with the activated sludge process, or its intensification, as long as my distinguished co-author. Activated sludge has been a part of my career for only the past 12 years, and it isn't how it started out. My background in the wastewater industry dates back to 2002 when I began graduate studies at École Polytechnique de Montréal. My topic of research was nitrification in municipal aerated lagoons. In fact, we were trying to figure out how to get the municipal lagoons in Québec to nitrify during the critical early summer to late fall period when effluent ammonia discharges were deemed to be harmful to the receiving water environment. A classic case of trying to get the existing infrastructure to do more where the undesirable alternative was replacement of the lagoons with activated sludge plants.

Lagoons are the ultimate "extensive" treatment technology requiring vast land surfaces, per population equivalent served, as compared to the relatively "intensive" activated sludge process. But extensive treatment has much to recommend itself. By relying on nature to provide mixing and oxygenation, and with very few mechanical requirements, it is arguably the most resilient and sustainable of treatment strategies. And yet my career, like the wastewater industry as a whole, has been headed in the opposite direction. I guess the continuing trend towards urbanization and a global population of over 7 billion people have something to do with this. Here I am 17 years later writing a book on how we should be further intensifying the activated sludge process. How things have changed.

My motivations for this book include trying to, if possible, once and for all summarize my thoughts on the process design of hybrid biofilm/suspended growth systems, the seeding effect, and the special properties of ammonia-limited *vs.* oxygen-limited biofilms. I think there is a gap in our collective knowledge on these topics. I have heard too many times practioners talk about what they believe or don't believe concerning seeding effect and hybrid processes. It shouldn't be a matter of belief so much as understanding and examining the assumptions that underpin the claims around process intensification in hybrid systems. Hopefully, by putting these thoughts out there for the professional community, I might be able to pursuade one or two people to see things my way. Or, if not, maybe solicit some interesting rebuttal that could help in the evolution of my own thinking. Knowledge is a two-way street and I am just as happy to learn from this venture as to teach others.

I have brought to this book my background as a commercial process model developer, a process designer/consultant, and more recently as a process equipment developer. My biases derive from these experiences. Working for a commercial model software developer showed me the power, and limitations, of models to predict full-scale behavior. I learned to respect model complexity for its descriptive power, but not always its usefulness in practice. Familiarity perhaps bred some measure of contempt in me, or at least fatigue, and I left my role as a model developer to join a global engineering firm. I wanted to be closer to "practice". But models were unavoidable in that environment too. In almost every project I worked on, the process model was highly valued and central to the decision making. My appreciation for this was heightened, and my faith in the value of good process models fully restored, when I worked on a plant commissioning in which there was no process model at all. Nobody had a clue what was going on.

Working for a global engineering and consulting firm is where I was first exposed to hybrid processes and became caught up in the industry enthusiasm for process intensification. It was also where I was first exposed to the Membrane Aerated Biofilm Reactor (MABR) technology. I was reviewing an equipment vendor's proposal for a ZeeLung MABR on behalf of a municipal utility. It seemed intriguing as an ultra high efficiency aeration device, but the vendor was pitching it as an alternative to conventional biofilm technologies. Knowing nothing about MABR, I was intrigued but struggled to understand how it would provide meaningful benefit to the client. In the industry jargon, I was struggling to understand the "value proposition".

I guess not understanding is really uncomfortable for me because, not long after, I joined the ZeeLung team at GE Water, now SUEZ Water Technologies & Solutions. In addition to getting me first hand exposure to MABR, its state of development and what it can do, it also slashed my commuting time in half. This is saying something in the traffic-choked roads of the Greater Toronto Area. In addition to my R&D duties, I now had the task of being the one to explain the ZeeLung "value proposition".

When speaking to people about the benefits of MABR, the least intuitive claims seem to be those relating to process intensification. That is to say, the opportunity to increase treatment capacity within the existing bioreactor tank volumes. Seeding effect is often at the heart of these claims. Explaining the benefits of, but also the limits to, the seeding effect has been a part of my daily functions since joining the ZeeLung team. This book represents an attempt to formalize my thinking about MABR for process intensification, including the role of seeding effect, and how it impacts real-world process design. While the impetus for this book is the development of hybrid MABR processes, the discussion, theoretical framework, and process modeling applies to other forms of integrated fixed-film/activated sludge (IFAS) including MBBR/AS and fixed-media/AS.

Dwight Houweling, Hamilton, Ontario, 8 March 2019

1 The Traditional Activated Sludge Design SRT

This book explores intensification of the activated sludge process on the assumption that nitrification is a performance requirement, and that maintaining robust nitrification is a constraint on rated capacity. Capacity can be defined as the flow, load or population equivalent that can be treated. Rated capacity refers to how much is "allowed" to be treated, either by the environmental regulator or, if one exists, in a process performance warranty. Capacity is used as the fundamental metric of "quantity". While the trend in the industry is towards ever more stringent effluent "quality" requirements at the discharge end of the plant, it has rather unsurprisingly led to derating of the "quantity" that can be received at the front end.

A classic example is a non-nitrifying conventional activated sludge (CAS) plant that receives a new effluent permit requiring it to remove ammonia. The resulting upgrade to a nitrifying process may derate the plant capacity by a factor of three. Or else a nitrifying plant that receives a new effluent permit requiring it to remove total nitrogen (TN). The resulting upgrade to a biological nutrient removal (BNR) process likely derates capacity by 30% or more.

In each of these cases, the loss of capacity may be compensated for by the construction of new bioreactor tanks and clarifiers. This is costly. At the same time, increasing populations are creating a separate driver for not just maintaining, but increasing treatment capacities. This means even more tanks and clarifiers. Process intensification offers an alternative roadmap to increase or restore treatment capacity within the existing treatment infrastructure.

In this book, we will assume that sludge retention time (SRT) under winter conditions is the most important metric governing nitrification. So a discussion of SRT is where this chapter starts and, indeed, it will be the focus of much of this book. This may reflect a "cold weather" bias and we acknowledge that there are many regions of the world where design is not governed by winter conditions. How the insights and conclusions provided in this book may apply in warmer weather regions will be discussed towards the end of this chapter.

The purpose of this chapter is to establish the basis for design of the conventional activated sludge process. More specifically it will provide an overview of:

- The basis of traditional design in theory and operational experience.
- How safety factors are applied to accommodate a range of factors including management of peak loading events.
- The opportunity cost of operating at extended SRTs.

1.1 THE MINIMUM SRT

1.1.1 NITRIFIER WASHOUT

Design of nitrifying activated sludge processes is governed by the need to retain the mixed liquor sludge for a sufficient sludge retention time (SRT) so that the nitrifying organisms can convert the influent ammonia into nitrate. At very low SRTs, there are not sufficient nitrifiers, if any at all, because they are being wasted faster than they can grow. We call this condition washout. At what is sometimes called the critical or washout SRT, the system is right at the transition between washout and nitrification. Full nitrification is a term that is used when most of the influent ammonia is being nitrified and effluent ammonia is low, less than 1 mg N/L. Full nitrification, like washout, is a term used for convenience, there are no precise definitions.

It has been one of the axioms of activated sludge process design that the transition from washout to full nitrification is almost immediate. This is illustrated in the graph of effluent ammonia *vs.* SRT in Figure 1.1, the so-called washout curves. From these curves we see that a washout SRT of just less than 2 days is expected at 20°C, whereas a washout SRT is closer to 6 days at 5°C. The same information is presented in a different form in the graph of Minimum SRT *vs.* temperature for the case of a target effluent ammonia concentration, S_{NHx}, of 5 mg N/L. The Minimum SRT is defined as the SRT required to achieve a given effluent ammonia objective. Figure 1.1 shows how the minimum SRT is extended if the effluent objective decreases to 1 or 0.5 mg N/L. The lesson being that achieving more stringent effluent objectives requires operating at longer SRTs.

Design SRT is yet another term that is used to describe the SRT that is selected as the basis for sizing bioreactor tanks and clarifiers. It should be longer than the Minimum SRT because it includes some safety factor. As will be discussed in Section 1.3, a Design SRT of 10 to 15 days is common for wastewater temperatures operating in the range of 10°C. This is true for plants that need to meet stringent, as well as moderate, effluent ammonia requirements. This leads to the following two, seemingly contradictory, observations from Figure 1.1:

1. Theory indicates that a much *shorter* Minimum SRT is required to meet a moderate (5 mg N/L) effluent limit than what is being applied in practice.
2. Theory also indicates that a much *longer* Minimum SRT is required to meet stringent (1 mg N/L or less) effluent limit than what is being applied in practice.

For the first point, safety factors can explain the need to operate at longer SRTs than is indicated from the washout curve. This will be discussed in greater detail in Section 1.2.1. The second point cannot be explained by safety factor, however. It relates to the assumptions underlying the curves presented in Figure 1.1, namely the assumption of a completely mixed bioreactor. This assumption greatly facilitates the development of design curves and equations, but it is not representative of reality in most cases. In practice, bioreactors are designed to have some some degree of plug-flow behavior and this considerably improves their ability to achieve low effluent ammonia concentrations, *i.e.* polishing.

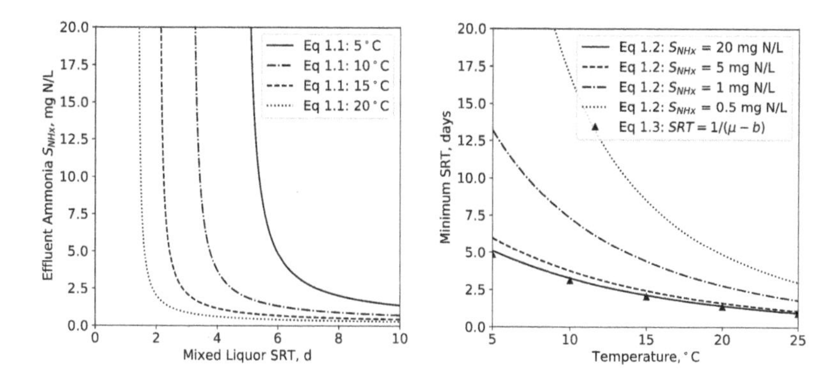

Figure 1.1 Washout curves based on Eq. 1.1 and Minimum SRT based on Eq. 1.2

While the curves presented in Figure 1.1 are instructive, their lessons need to be applied with caution. In addition to the points noted above, there can be other factors that need to be taken into account. For example, the transition from full nitrification to washout can be more gradual than is illustrated in Figure 1.1. Also, Figure 1.1 does not account for the separate contribution to nitrification of ammonia oxidizing bacteria (AOBs) and nitrite oxidizing bacteria (NOBs). Normally, these organisms are expected to work together and we account for their activity using a single set of biokinetic parameters for nitrifying organisms, X_{Nit}, or, as is the approach in this book, we assume that the first step of nitrification is the rate limiting one, and we predict nitrification behavior based on the biokinetic parameters of AOBs, X_{AOBs}. But it is well known that NOB washout tends to occur before AOB washout and elevated concentrations of effluent nitrite, sometimes called "nitrite lock" may be observed when operating at low SRTs. Owing to the toxicity of nitrite, both to the activated sludge biomass and for the effluent receiving water environment, this is an undesirable condition.

1.1.2 DESIGN EQUATIONS

The basis for the curves presented in Figure 1.1 are Equations 1.1, 1.2 and 1.3 as presented below. The use of these equations is well established and their derivation has been presented in many textbooks including Metcalf & Eddy and Grady *et al.* [1, 10].

Equations 1.1 and 1.2 assume mixed-order kinetics to describe the growth of nitrifying organisms: first order with respect to biomass and Monod kinetics with respect to substrate. First order kinetics are assumed for biomass decay. The equations are developed by solving the mass balances on the organism population, X_{AOB}, and ammonia substrate, S_{NHx}, in a complete-mix activated sludge process (CMAS). Note that in this book, AOBs will act as a surrogate for all nitrifying organisms under the assumption that their growth is rate limiting to nitrification. This approach is similar to assuming that nitrification is a single step process which lumps AOBs and NOBs into a single organism group, X_{Nit}, but allows the biokinetic parameters to be aligned with those used in simulation software, which tend to use two-step nitrification models. A more in depth discussion of these equations, and how they should be applied to hybrid biofilm systems, will be presented in Chapter 4.

$$S_{NHx} = \frac{K_N(1+bSRT)}{SRT(\mu - b) - 1} \tag{1.1}$$

$$SRT = \frac{K_N + S_{NHx}}{S_{NHx}(\mu - b) - K_N b} \tag{1.2}$$

Equation 1.3 presents an interesting simplification to Equation 1.2 in that it requires no knowledge of effluent ammonia concentration. In fact, it calculates the SRT as simply equivalent to the reciprocal of the growth, μ, minus decay, b. As can be seen from Figure 1.1, the SRT calculated from Equation 1.4 is only slightly lower than the minimum SRT calculated with Equation 1.2 to achieve effluent ammonia concentrations of 5 and 20 mg N/L, respectively. In fact, Equations 1.2 and 1.3 converge as effluent ammonia, S_{NHx}, increases.

$$SRT = \frac{1}{\mu - b} \tag{1.3}$$

The concentration of nitrifying organisms, X_{AOB}, in the bioreactor is calculated based on ammonia removal, $(S_{NHx,0} - S_{NHx})$, and the growth yield corrected for decay, $Y/(1+bSRT)$. In addition, this equation includes the ratio of SRT/HRT to account for the decoupling of SRT and HRT in the activated sludge process. In contrast, for a lagoon type process, this term would be unnecessary since $SRT = HRT$. If one assumes an SRT of 10 days and an HRT of 6 hours, then the ratio of SRT/HRT is 40. This is indicative of the level or process intensity of activated sludge as compared to lagoon systems.

$$X_{AOB} = \frac{SRT}{HRT}\frac{Y(S_{NHx,0} - S_{NHx})}{1+bSRT} \tag{1.4}$$

In Equations 1.1 to 1.4, the specific growth rate of nitrifying organisms at a given temperature is calculated based on the reference temperature of 20°C.

$$\mu = \mu_{20C}\theta_{\mu}^{T-20} \tag{1.5}$$

In the same manner, the specific decay rate of nitrifying organisms at a given temperature is calculated based on the reference temperature of 20°C.

$$b = b_{20C}\theta_b^{T-20} \tag{1.6}$$

Table 1.1 lists and defines the calculated or operationally defined parameters for Equations 1.1 to 1.4. Influent ammonia concentration, $S_{NHx,0}$, SRT and HRT may be primarily thought of as operationally defined parameters. They can be fixed by the operator or in design or else are a consequence of influent wastewater characteristics. Bioreactor nitrifying organisms, X_{AOB} is a calculated parameter. SRT and effluent

Table 1.1
Calculated or Operationally Defined Parameters for Equations 1.1, 1.2 and 1.3

Parameter	Units	Description
$S_{NHx,0}$	mg N/L	Influent ammonia concentration
S_{NHx}	mg N/L	Bioreactor ammonia concentration which is equivalent to effluent ammonia for a complete mix system
X_{AOB}	mg N/L	Concentration of nitrifying organisms in the bioreactor. In this case, AOBs are assumed to be rate limiting to nitrification and act as a surrogate for all nitrifiers.
SRT	days	Sludge retention time
HRT	days	Hydraulic retention time

Table 1.2
Biokinetic Parameters Used to Develop Figure 1.1

Parameter	Value	Units	Description
μ_{20C}	0.9	mg COD/mg N/d	Specific growth rate of nitrifiers at reference temperature of 20°C
b_{20C}	0.17	mg COD/mg COD/d	Specific decay rate of nitrifying organisms at reference temperature of 20°C
K_N	0.7	mg N/L	Half saturation concentration for ammonia
θ_μ	1.072	-	Arrhenius coefficient for temperature dependence of growth rate
θ_b	1.029	-	Arrhenius coefficient for temperature dependence of decay rate
Y	0.15	mg COD/mg N	Growth yield of AOBs for conversion of ammonia to nitrite

ammonia, S_{NHx}, alternate as either operationally defined or calculated parameters depending on whether Equation 1.1 or 1.2 is applied.

The biokinetic parameters listed in Table 1.2 served as the basis of developing the curves in Figure 1.1 and will be used throughout this book. Nitrifying biomass is measured in units of chemical oxygen demand (COD), as opposed to volatile suspended solids (VSS), to align with the approach used in simulation software. While a range of values may be found in the literature for AOB biokinetic parameters, the values listed in Table 1.2 are representative of what is being used in at least three of the major commercially available simulation sofware packages.

1.2 THE DESIGN SRT

1.2.1 SAFETY FACTORS

Equations 1.1 and 1.2 provide what is called a "kinetic basis" for calculating the minimum SRT to meet a target effluent ammonia concentration. Good engineering judgment is required in applying these equations because they do not account for factors such as plug-flow conditions in the bioreactor and influent loading dynamics. Typically a safety factor, SF, will be applied to the calculated "minimum" SRT from Equation 1.2 as follows:

$$SRT_{Design} = SF \times SRT_{min}$$

According to Metcalf & Eddy, the justification for using a safety factor is to (1) allow flexibility for operational variations in controlling the SRT, and (2) to provide additional nitrifying bacteria to handle peak TKN loadings [1]. Although the relative impacts of these two factors may be lumped together into a single factor, given the disparity in peaking factors between small and large scale plants, it is appropriate in many cases to treat them separately. This is the approach taken in the German ATV Design Guidelines which are discussed in Section 1.2.1.

A somewhat different approach to safety factor is presented by Grady *et al.* who identify, in addition to (1) influent peak loading, the need to also account for (2) uncertainty in kinetic parameters, influent characteristics, natural variability in the microbial community, and other factors, as well as (3) dissolved oxygen limitations [10]. Grady *et al.* note that application of several safety factors in a multiplicative manner can lead to systems that are grossly overdesigned. To correct for this, their recommendation is to apply the three safety factors to the SRT calculated from Equation 1.3 *i.e.* the washout SRT.

Controlling the SRT

Controlling SRT in a full-scale plant requires first of all to control wasting on a "volume per day" basis. The reliability of this is subject to the quality of waste activated sludge (WAS) flow metering and may be impacted by the schedule and priorities of the biosolids handling facility, which must receive the wasted sludge. But it is wasting on a "mass of solids per day" basis that is fundamental to determining the SRT. This introduces additional uncertainty around waste activated sludge solids content in "all of the sludge" leaving the plant on a daily basis. Admittedly, measuring the solids content of a waste activated sludge sample in the lab is a simple procedure. Obtaining a representative sample of "all of the sludge" leaving the plant, however, is not.

Peak TKN Loadings

Providing added safety factor to handle peak TKN loadings is necessary because Equations 1.1 and 1.2 are derived for steady-state conditions *i.e.* changes with respect to time $\frac{d}{dt}$ are assumed to be nil. So by very definition, Equations 1.1 and 1.2 do not account for the loading variations that occur on a diurnal, weekly and seasonal basis. In practice, we see that effluent ammonia can be highly sensitive to loading variations. One reason for this is because nitrifying organisms, unlike heterotrophs, lack the ability to store substrate during peak loading "feast" conditions. As a result, during these events, the increase in ammonia load "over and above" the average loading may pass directly through into the effluent. This is commonly referred to as effluent breakthrough.

1.2.2 RATIO OF NITRIFIER ACTIVITY TO AMMONIA LOAD

The effectiveness of applying a safety factor to manage load peaks, and avoid effluent ammonia breakthrough, is illustrated in Figure 1.2. The strategy consists essentially of using the SRT to multiply the mass of nitrifying organisms, and therefore the maximum potential nitrification activity in the bioreactor, "over and above" what is required to simply treat the average loading condition.

Mass of Nitrifiers

The mass of nitrifying organisms can be calculated from the concentration of nitrifying organisms, X_{AOB}, defined in Equation 1.4, and the bioreactor volume, V, as follows:

$$M_{AOB} = X_{AOB} \times V \tag{1.7}$$

Maximum Potential Nitrification Activity

The maximum potential nitrification activity, $A_{Nit,max}$, is simply the mass of nitrifiers, M_{AOB}, multiplied by the specific substrate utilization rate, μ/Y:

$$A_{Nit,max} = \frac{M_{AOB}\mu}{Y} \tag{1.8}$$

This relationship assumes ammonia is kinetically non-limiting, which is why it is referred to as the "maximum" potential nitrification activity. It is analogous to the activity you would measure at the front end of the bioreactor where ammonia concentrations are high. Or if you took a sample of mixed liquor to the laboratory, spiked it with 20 mg N/L of ammonia, and measured the rate of ammonia removal in a batch test. This level of activity would not be observed under the ammonia-limited conditions that occur at the downstream end of activated sludge bioreactors. But that is not the point. Under peak loading events, when higher ammonia concentrations break through into the downstream end of the bioreactor, nitrifying organisms that were formerly ammonia-limited are able to increase their nitrification rate to match the new load.

The ratio of maximum potential nitrification activity, $A_{Nit,max}$, to the average influent load, *Load*, is defined as follows and has units of kg N/d per kg N/d:

$$R_{Nit,max/L} = \frac{A_{Nit,max}}{Load} \tag{1.9}$$

The ratio $R_{Nit,max/L}$ has the potential to range from less than one, to two or even three. Higher values of $R_{Nit,max/L}$ indicate greater ability to treat peak loading events. It is important to note, however, that $R_{Nit,max/L} > 1$ does not mean that the nitrifiers are removing more than what is available in the influent load. It only means that,

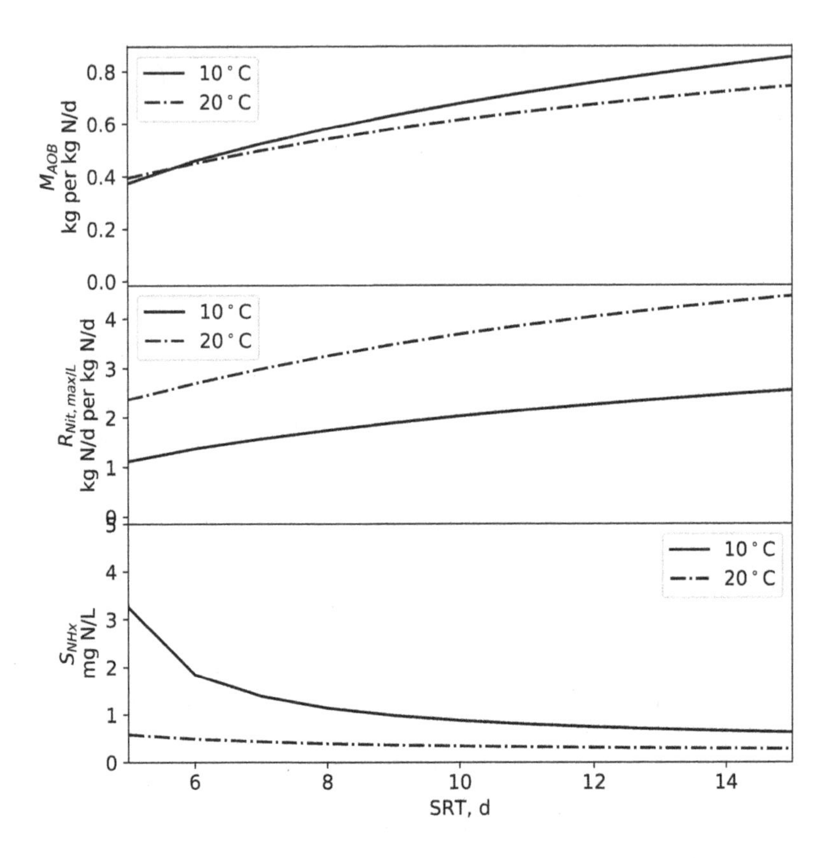

Figure 1.2 Ratio of nitrifying organisms inventory (AOBs) to ammonia loading as a function of mixed liquor SRT at 10°C and 20°C

when peak loading events occur, the nitrifiers have the potential to "rise to the occasion" and remove more ammonia than they would under average loading conditions.

Figure 1.2 shows that at 10°C and an SRT of 5 days, the effluent ammonia is approximately 3 mg N/L and $R_{Nit,max/L}$ is approximately 1 kg NH_4N/d of potential nitrifier activity per kg NH_4N/d of average day influent loading. So even if an effluent ammonia concentration of 3 mg N/L were acceptable, with $R_{Nit,max/L} = 1$, the ability to treat any short-term peak in the influent ammonia load is effectively nil. This is hardly a desirable operating point. At an SRT of 10 days, however, the potential activity is almost double the average loading, *i.e.* $R_{Nit,max/L} \approx 2$. One might interpret from this that, at a temperature of 10°C, an operating SRT of 10 days is appropriate for treating an influent load with a peaking factor up to 2. As discussed in Section 1.2.1, there are many factors that need to be accounted for in the application of safety factor. But the principle applies that higher SRTs are an effective strategy for avoiding effluent ammonia breakthrough during peak loading events.

The key points of interest from Figure 1.2 are summarized as follows:

- The effluent ammonia concentration is essentially the same for all SRTs above 5 days at 20°C, and above 10 days at 10°C. On its own, looking at effluent quality in this figure indicates no meaningful benefit to operating above these minimum SRTs.
- The ratio of potential activity to average loading increases by almost 50% between SRTs of 10 and 15 days, at 10°C, and by almost 100% between 10 and 20 days, at 20°C. This increases the ability to treat short-term loading peaks and explains the benefit of applying safety factors to the minimum SRT.
- The increase in nitrifier inventory as a function of SRT is not linear because biomass decay also increases as a function of SRT.
- In almost all cases, except the case where effluent ammonia is above 3 mg N/L, the inventory of nitrifying organisms is higher at 10°C than it is at 20°C. This is because decay is more rapid at higher temperatures.

While the curves presented in Figure 1.2 were developed using process modeling software, they could just as easily have been developed from Equations 1.1 and 1.4. The use of process modeling software, and its advantages, will be discussed in greater detail in Chapter 5. In addition, the effectiveness of hybrid processes in providing an alternative means to increase the ratio of maximum potential activity to average influent load will be explored in Chapter 6.

1.2.3 APPLICATION IN DESIGN

To develop realistic strategies for intensification of the activated sludge process, the role of SRT in design needs to be properly understood. That is to say, to make bioreactors and clarifiers smaller we first need to understand how using the traditional design SRT leads to larger bioreactors. This SRT-based design procedure is described in the paragraphs that follow. In Section 1.3.2, an alternative method will be discussed that relates nitrification to volumetric loading rates based. This empirical method does not require any knowledge of operating SRT whatsoever. As will be shown, however, the empirical approach leads to similar outcomes as the kinetically based, SRT-based design procedure.

Minimum Aerobic SRT

The minimum SRT is identified using washout curves derived from Equation 1.1. The temperature and target effluent ammonia concentrations are key parameters that impact the minimum SRT that is derived from this curve. It is recognized that the minimum SRT calculated in this manner assumes fully aerobic conditions.

Effluent nutrient requirements may require inclusion of unaerated zones to promote denitrification and/or enhanced biological phosphorus removal (EBPR). The presence of unaerated zones would be accounted for by assigning the minimum SRT from Equation 1.2, or Equation 1.3, to the minimum aerobic SRT. The minimum total

SRT would then be calculated as the minimum aerobic SRT divided by the aerobic mass fraction of the mixed liquor.

Dissolved oxygen limitations in the bioreactor would be accounted for by reducing the specific growth rate of nitrifying organisms μ proportional to a half saturation coefficient for dissolved oxygen K_{DO} and assuming a *Monod* relationship $\frac{DO}{K_{DO}+DO}$.

Safety Factor

A safety factor is applied to the minimum SRT to account for influent load variability, uncertainty in nitrifier kinetics, and to provide operational flexibility in managing sludge wastage. This safety factor may be applied explicitly, as in the ATV design guidelines, Section 1.3.1, or implicitly as in the Ontario government design guidelines, Section 1.3.3.

Sludge Production and Inventory

The sludge production is calculated as a function of the influent BOD, or COD, and TSS load and the sludge yield. The sludge yield is defined as the mass of sludge produced per day relative to the incoming load. Sludge yields can vary due to the site specific influent wastewater characteristics, industrial load contributions and other factors. Sludge yields also vary according to a well defined relationship with SRT, where lower sludge yields are observed at longer SRTs. Much of the uncertainty in process design lies around quantifying sludge production. The source of this uncertainty typically is due to inaccurate quantification of variable influents loads and the relative fractions of biodegradable and unbiodegradable material in the TSS.

The sludge inventory is calculated as the sludge production multiplied by the SRT. This is a straightforward calculation but any uncertainty in error in assumed sludge production, or SRT, will be reflected in the calculated sludge inventory.

Bioreactor Volume

The bioreactor volume is simply calculated as the sludge inventory divided by some target mixed liquor concentration. A target MLSS of 3,000-4,000 mg/L is very typical, however, there is no practical limitation to holding mixed liquor in the bioreactor at concentrations of 10,000 mg/L or higher. At very high MLSS, oxygen transfer efficiency (OTE) declines significantly, and in the smaller bioreactor volumes one expects with process intensification, it may not be possible to provide sufficient aeration to meet the oxygen demand. A target MLSS in the range of 3,000-4,000 mg/L usually means solids loading rate to the secondary clarifier is the guiding design criteria.

Clarifier Sizing

The two basic design parameters for secondary clarifiers are the surface overflow rate (SOR) and the solids loading rate (SLR). The SOR is a function of the plant influent flows whereas the SLR combines the effect of influent flow, return activated sludge (RAS) flow and the mixed liquor suspended solids (MLSS) in the bioreactor.

Increasing the size of the secondary clarifier can enable operation at higher bioreactor MLSS. That is to say that a larger clarifier can lead to a design for a smaller bioreactor. A design MLSS of 3,000-4,000 mg/L is typically thought of as providing an optimal compromise between the sizing, and cost, of bioreactors as compared to clarifiers.

Aeration System Design

The aeration system must provide the oxygen required to maintain aerobic conditions in the bioreactor while meeting the carbonaceous and nitrogenous oxygen demand of the biomass. The relative contribution of these two to the total oxygen demand tends to be about even, but this can vary based on the carbon to nitrogen ratio in the influent wastewater, and the operating SRT.

Typical design would involve providing the oxygen requirement and the bioreactor geometry to an aeration equipment supplier. Estimates would be made as to the relative distribution of oxygen demand throughout the tank, placing a higher proportion of the demand at the upstream end. The airflow per diffuser for the tank would then be calculated assuming an oxygen transfer efficiency (OTE), typically about 10 to 20%. The main source of uncertainty in this procedure being the identification of the mass transfer resistance to oxygen transfer under the site specific conditions, the so-called α factor.

For conventional activated sludge, the aeration system design would not typically impact choices made with respect to bioreactor or clarifier sizing. At high temperatures or with high strength wastewater, however, this might not be true. In these cases, it is necessary to increase the bioreactor size to "dilute" the oxygen requirement, kg O_2/d, over more volume, m^3. The key metric which accounts for these two parameters is the oxygen utilization rate (OUR) which is typically expressed in mg O_2/L/h. Designing for OUR's greater than 90-100 mg/L/h becomes impractical with conventional aeration technologies. This is a point to keep in mind as we further investigate the topic of activated sludge intensification.

1.3 REGULATORY GUIDELINES FOR NITRIFYING ACTIVATED SLUDGE

1.3.1 ATV GUIDELINES

The German government produces a set of standards and guidelines commonly referred to as the ATV guidelines. The ATV guidelines are very influential within the industry being used not only in Germany, but many other countries in Europe as well. According to the ATV guidelines, the basis for tank sizing is the design SRT, a calculated sludge production, and maximum acceptable clarifier loading rates [25].

The design SRT is derived from a combination of kinetic parameters for nitrifying organisms and a safety factor. The purpose of this safety factor is to take into account the following:

- variability in the growth rate of nitrifying organisms, $\mu - b$, caused by inhibitory substances in the influent wastewater, short-term temperature variations and/or pH shifts,

- the effluent ammonia objectives, S_{NHx}, and
- influent nitrogen loading variation, *i.e.* peaking factor.

Since lower capacity treatment works support smaller populations and have smaller sewersheds, they generate more diurnal and seasonal load variability. Smaller treatment works therefore have higher peaking factors and, according to the above rationale, need to be designed with a higher safety factor. Lower capacity treatment works also tend to benefit from less attention and staffing, and so applying higher safety factors to smaller plants helps in this regard too.

1.3.2 TEN STATES STANDARD

A series of recommended standards for the design of wastewater treatment facilities is published by the Wastewater Committee of the Great Lakes - Upper Mississippi River commonly referred to as the Ten States Standards [22]. Member states and provinces on this committee include the U.S. states of Illinois, Indiana, Iowa, Michigan, Minnesota, Missouri, New York, Ohio, Pennsylvania and Wisconsin and the Canadian province of Ontario. The Ten States Standards have an important influence on the design guidelines imposed by regulators in these jurisdictions and others.

Loading Based Guidelines

In contrast to the ATV design guidelines, it is interesting to note that the Ten States Standards provide no guidance on minimum SRT when designing conventional activated sludge. The approach is much more empirical and focusses on maximum loadings of BOD per tank volume. The Ten States Standards for process design of conventional activated sludge are presented in Table 1.3. The key number in this table is the maximum loading of 15 lbs BOD/d per 1000 ft^3 for a single stage nitrification process.

Relationship Between Loading and SRT

The curves presented in Figure 1.3 demonstrate that a loading rate of 15 lbs BOD_5/d per 1,000 ft^3 of bioreactor volume corresponds to an SRT greater than 15 days. Variables that impact this relationship include influent wastewater characteristics and target mixed liquor suspended solids. In the case of Figure 1.3, influent wastewater characteristics have been characterized in terms of the TSS/BOD ratio of the influent which has been varied from 0.9 to 1. A higher TSS/BOD ratio would be indicative of a greater contribution of non-biodegradable suspended solids (organic or inorganic) in the influent feed. The contribution of non-biodegradable suspended solids in the influent feed is important because it contributes to sludge yield at a ratio of 1:1. This means that 100% of every lb (or kg) of influent non-biodegradable suspended solids contributes to sludge production. In contrast, the BOD in the feed contributes at a much lower ratio, less than 0.5:1, because a portion of it is oxidized to CO_2 and H_2O.

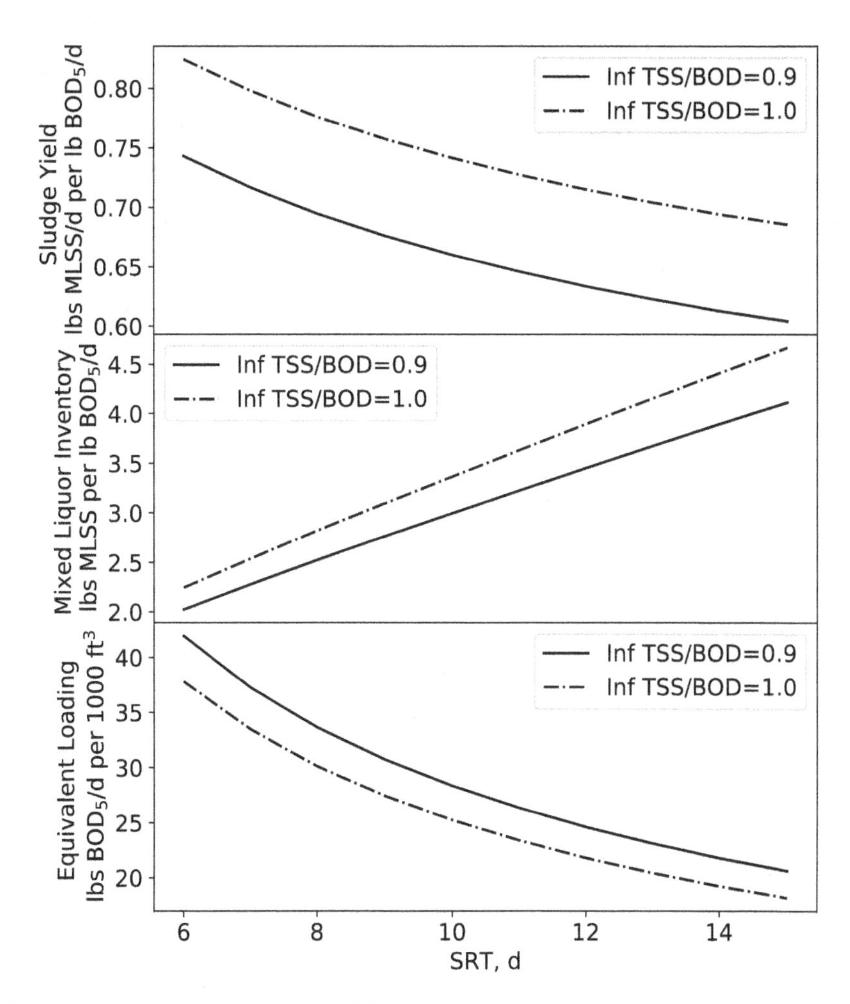

Figure 1.3 Relationship between SRT and allowable loading rates based on process model simulation assuming target MLSS = 3,000 mg/L and 20°C

The target MLSS used in design can vary significantly and is governed by the sizing of the secondary clarifier. A larger secondary clarifier will enable design of the bioreactors at a higher MLSS. Table 1.3 presents a range of 3,000 to 5,000 mg/L. Design MLSS of 3,000 mg/L to 3,500 mg/L are probably most typical with higher operating values being allowed for under the maximum month loading condition.

1.3.3 ONTARIO GOVERNMENT GUIDELINES

It is interesting to observe how the Ten States Standards are applied at the regulatory level by one of its members states, actually a province. The Government of Ontario publishes its "Design Guidelines for Sewage Works" which are summarized in Table 1.4 for the purpose of comparison with Ten States Standards, summarized in Table 1.3 [22, 21].

Note that the Ontario Government Guidelines retain the same loading limits as the Ten States Standards, but also identify acceptable ranges of operating SRT. The target loading rate of 15 lbs BOD_5/d per 1000 ft^3, or 0.24 kg BOD_5/d per m^3, identified in the Ten States Standard reappears as the maximum loading rate for the Extended Aeration process in Table 1.4. Consistent with Figure 1.3, an SRT of greater than 15 days is identified for this loading condition. Extended aeration design criteria, SRT>15 days, tend to be applied for smaller capacity treatment works, where peaking factors are higher and there is less attention from operational staff.

For conventional nitrifying activated sludge, a more permissive loading rate of 310 to 720 g BOD_5/m^3/d is allowed for, corresponding to an SRT requirement of 10 days or greater. Compared to the extended aeration design criteria, the lower design SRT of 10 days sees application for larger capacity treatment works where peaking factors are lower, and there is more attention from operational staff.

1.3.4 GUIDELINES FOR WARMER REGIONS

The above guidelines were developed in a context where the minimum SRT for nitrification in winter is significantly greater than it is in summer. As a result, cold weather conditions govern the sizing of the bioreactor and clarifier. In warmer regions of the world, the governing condition for bioreactor sizing may in fact be summer conditions when high temperatures contribute to higher respiration rates in the mixed liquor, and higher resistance to oxygen mass transfer.

When the aeration system design governs the sizing of the bioreactor, the concern is likely not that nitrifying bacteria will be washed out of the system, but rather that they will not be provided sufficient oxygen to nitrify effectively. In these cases, the concept of safety factor, discussed in Section 1.2.1, is still useful. Applying a safety factor increases the ratio of maximum potential nitrification activity to influent loading $R_{Nit,max/L}$. In addition to managing influent load ammonia peaks, this provides additional nitrifying organisms to compensate for the loss of activity that may occur during periods of dissolved oxygen limitation, the so-called DO sag.

1.4 THE OPPORTUNITY COST OF LONG SRTS

The preceding sections argue for the benefits of a designing with a high safety factor and long SRTs and indeed, in practice, this is often the guiding design philosophy. For example, the Ontario government guidelines specify a minimum SRT of 10 days at temperatures of 5°C but they also recommend more than 15 days SRT for so-called "Extended Aeration" processes. The design of Extended Aeration processes is recommended in lower capacity applications where influent variability may be greater, and the level of operational support lower.

But, there is also an opportunity cost, or lost opportunity, associated with this. The most obvious being that long SRTs require larger bioreactors and clarifiers, and

Table 1.3
Permissible Aeration Tank Capacities and Loading According to Ten States Standards

Process	Aeration Tank Organic Loading lbs BOD_5/d/1000 ft^3 (kg BOD_5/d/m^3)	F/M ratio lbs BOD_5/d/lb MLVSS (kg BOD_5/d/kg MLVSS)	MLSS mg/L
Conventional Activated Sludge Step Aeration Complete Mix	40 (0.64)	0.2-0.5	1,000-3,000
Contact Stabilization	50 (0.80)	0.2-0.6	1,000-3,000
Extended Aeration Single Stage Nitrification	15 (0.24)	0.05-0.1	3,000-5,000

Table 1.4
Aeration System Design Parameters According to Ontario Government Standards

Process	Aeration Tank Organic Loading (kg BOD_5/d/m^3)	F/M ratio (kg BOD_5/d/kg MLVSS)	MLSS days	SRT
Conventional Activated Sludge without Nitrification	0.31-0.72	0.2-0.5	1,000-3,000	4-6
Conventional Activated Sludge with Nitrification	0.31-0.72	0.05-0.25	3,000-5,000	>4 at 20oC; >10 at 5oC
Extended Aeration	0.17-0.24	0.05-0.15	3,000-5,000	>15

the investment made in this infrastructure is money that cannot be invested in other places like transportation, education or health care. But besides the financial cost, there are additional factors that should be considered including operating cost, carbon footprint, and potential conflicts with resource recovery. This is discussed in the paragraphs that follow.

Higher Operating Cost

A higher safety factor requires more energy in operations because, at extended SRTs, there is more time for the mixed liquor organisms to decay. This decay is referred to as endogenous respiration and has an oxygen demand associated with it. At long SRTs, the endogenous oxygen demand can contribute very significantly to the aeration requirements of a plant. And given that aeration already accounts for more than 50% of a plant's energy cost at 10 days SRT, the added energy required to operate at SRTs much longer than this is worth considering.

It should also be considered that plant design typically targets a loading projection that is at least 20 years in the future. So until this loading is reached, the plant will operate in an underloaded state, in some cases a severely underloaded state. Aeration requirement for the biology may even be less than is required for bioreactor mixing, that is the energy required to keep the biomass in suspension. As a result, these plants need to overaerate their bioreactors: over and above the biological requirements. The energy efficiency of "mixing limited" plants, commonly expressed in kWh/m3 treated or kW/PE, is poor.

Energy Recovery

In addition to the increased aeration that is required at long SRTs, the biomass that is oxidized through endogenous respiration needs to be considered. Endogenous respiration in fact creates a lost opportunity for capturing the energy that is embedded in that biomass. Well established techniques for capturing this energy include anaerobic digestion and thermal oxidation (incineration) and it is well known that long SRT processes produce less quantity, but also lower quality of sludge for energy generation.

As the wastewater industry tries to shift from a paradigm where plants simply provide "pollution control" to one of "resource recovery facilities", the impact of long SRTs needs to be considered. In fact, shorter SRTs are what is required for plants to improve their energy balances: they offer the opportunity to both limit energy needs and increase energy production. This seems counter-intuitive because many wastewater treatment facilities are trying to minimize sludge production: sludge handling is costly. However, once investments in sludge valorization infrastructure has been made, this creates a new incentive to maximize sludge production.

Implications for Biological Nutrient Removal

The impact of operating SRT on biological nutrient removal is not straightforward, it depends a great deal on the process configuration and aeration strategies. On the one hand, it has been shown that very long SRT systems can, in some cases, achieve superior total nitrogen (TN) removal. This is possible by coupling the *oxidation* of carbon that occurs during endogenous respiration, to the *reduction* (denitrification) of nitrate to nitrogen gas. But the conditions required to achieve this are unique and cannot be applied in many plants.

For most plants, operating at long SRTs will increase the amount of ammonia that is released from the sludge. This occurs because, as the endogenous respiration of the sludge converts the carbon and hydrogen into carbon dioxide and water, the nitrogen is released as ammonia. For plants that have to meet an effluent total nitrogen (TN) requirement, operating at a long SRT increases the effective nitrogen load that needs to be removed. And the same logic applies to total phosphorus (TP).

Implications for Nutrient Recovery

One of the easiest ways to recover nutrients from wastewater is in the sludge. As a result, any process that minimizes sludge production limits the options for nutrient recovery. Other technologies exist which can recover nutrients in the liquid stream, but they all face the same fundamental fact that it is easier to recover nutrients from a concentrated stream than a dilute one. As a general rule, by increasing bioreactor volumes, and limiting sludge production, extended SRTs contribute to this "dilution" effect. Extended SRTs thereby limit the potential to recover nutrients from wastewater.

1.5 WHAT DOES INTENSIFICATION MEAN?

1.5.1 LOWERING MIXED LIQUOR SRT

The goal of intensifying the activated sludge process is treating more flow in less volume. As will be discussed in the next chapter, hybrid biofilm/suspended growth systems are not the only strategy for achieving this. In addition to hybrid systems, important strategies for intensifying the activated sludge process include membrane bioreactors (MBR), mixed liquor ballasting, granular activated sludge, step feed, and mixed liquor degassing. But of these options, only hybrid systems work on the basis of minimizing the mixed liquor SRT. And, as was discussed in the previous section, minimizing the design SRT is an important strategy for enabling resource recovery.

1.5.2 MAINTAINING SAFETY FACTOR

The goal of intensification should not be to reduce safety factor. A safety factor is critical for managing incoming load variations and providing the flexibility required to operate the plant. Operating the activated sludge process at a reduced safety factor is not really process intensification at all, it simply increases the risk of effluent compliance exceedences.

As was presented in Section 1.2.2 there is a rational basis for thinking about safety factor that, according to the ratio $R_{Nit,max/L}$, relates the maximum potential nitrification activity in the mixed liquor to the influent loading. This metric will be revisited in Chapters 4 and 6 to evaluate the intensification claims of hybrid processes operated above the washout SRT.

2 Media Supported Biofilms and the Activated Sludge Process

As discussed in the introductory chapter, the requirement to maintain a sufficient mass, or inventory, of nitrifying organisms in the bioreactor is the basis for selecting the design SRT. The reason for this is that SRT, together with influent loading, determine the sludge production and mass of sludge in the bioreactor. The mass of sludge, bioreactor volume and flow, in turn, determine the sludge loading to the secondary clarifiers. And since secondary clarifiers fail when sludge loading is too high, SRT effectively limits influent loading. So we find that nitrification requirements are the fundamental bottleneck on plant treatment capacity.

Introduction of biofilm support media, or carriers, that are fixed in the bioreactor using retention sieves, or other means, is one way to get around this bottleneck. The biomass inventory is increased without impacting sludge loading to the secondary clarifiers. Just like in activated sludge mixed liquor, this "biofilm biomass" can be composed of both heterotrophic and nitrifying organisms. Good design practice promotes the growth of "nitrifying biofilms" that remove a portion of the ammonia, and thereby decrease the effective load that needs to be treated by the mixed liquor. But in this type of hybrid process, is the required SRT for mixed liquor nitrification still the same as it would have been for a conventional process? If so, then what advantage does a hybrid process really offer?

This chapter provides an overview of hybrid biofilm/suspended growth technologies, also referred to as integrated fixed-film/activated sludge (IFAS) processes, their history and basis of design. More specifically, this chapter:

- Compares hybrid biofilm/suspended growth systems to alternative strategies for process intensification.
- Overviews IFAS design procedures which allow for a lower mixed liquor SRT than would otherwise be permissible in conventional activated sludge.
- Identifies different approaches to IFAS including Fixed-media IFAS, MBBR/AS IFAS and MABR/AS IFAS.
- Overviews how IFAS has evolved to better solve the factors which limit the intensity of the activated sludge process.

2.1 INTENSIFICATION STRATEGIES

When considering the potential of media supported biofilms to intensify the activated sludge process, it is helpful to ask a few questions. Firstly, what are the design

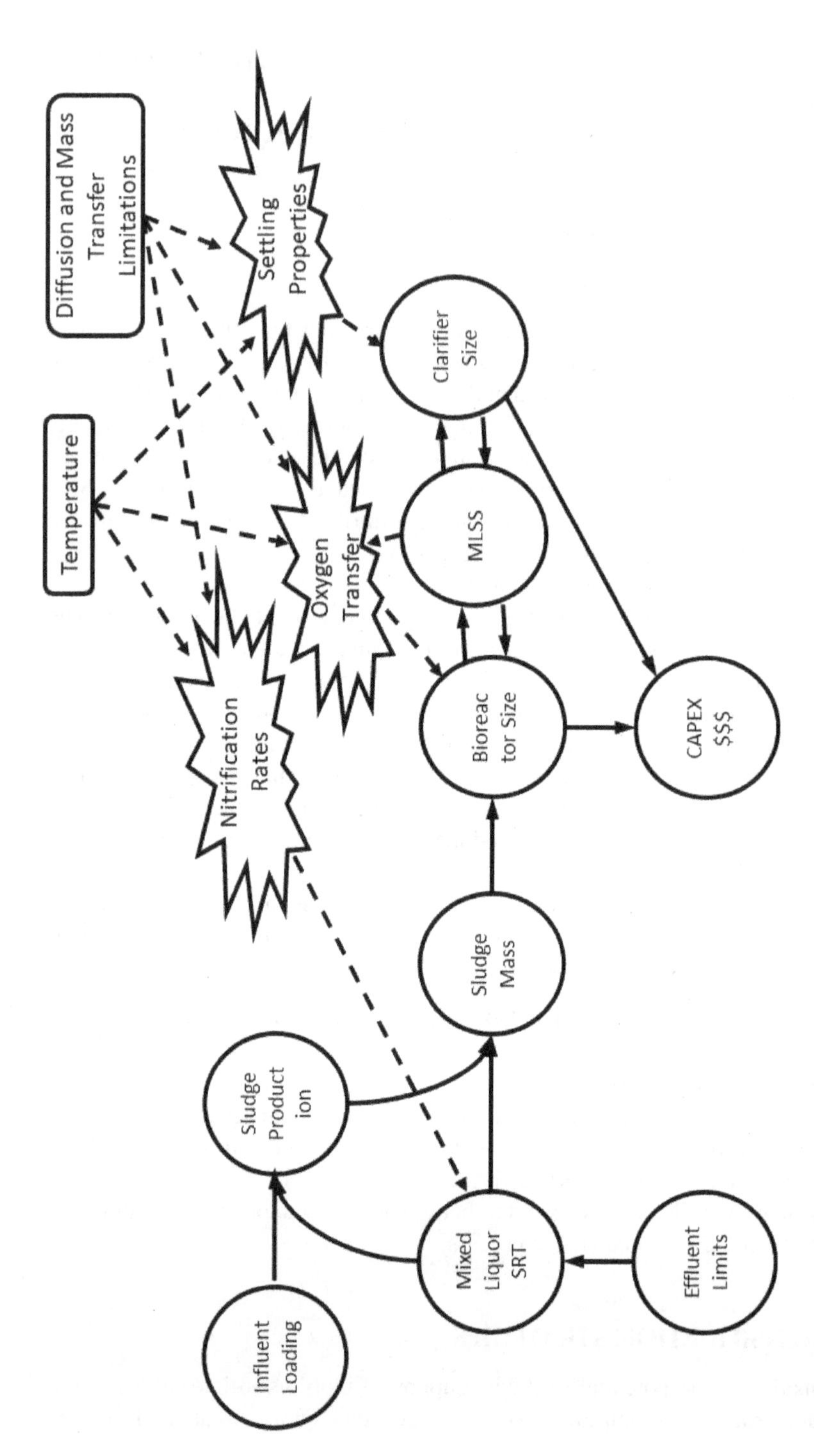

Figure 2.1 Schematic representation of conventional activated sludge design procedure including role of temperature and mass transfer limitations

procedures, and decisions made, for sizing activated sludge bioreactors and clarifiers? Secondly, what are the fundamental physical laws that constrain this process? And thirdly, what are the strategies to avoid these constraints which can result in smaller tank sizes, *i.e.* process intensification?

Figure 2.1 provides a schematic representation of how some of the key inputs and physical constraints to the activated sludge process contribute to bioreactor and clarifier sizing and, ultimately, the capital cost (CAPEX). These include influent flows and loads, temperature, and diffusion and mass transfer limitations. Although not a comprehensive overview by any means, Figure 2.1 provides useful insight into the role of *Mixed Liquor SRT*, relative to other factors such as *Temperature* and *Settling Properties*, in driving activated sludge *CAPEX*. Intensification of the activated sludge can be seen as an attempt to short-circuit, or "hack", this design procedure. For example, the clarifier can be replaced by a solids separation step that does not rely on *Settling Properties*. Or the *Settling Properties* of the activated sludge flocs can be enhanced using ballast or other means. Or, as is the case in a hybrid process, a portion of the *Sludge Mass* can be immobilized inside the bioreactor so that it does not impact clarifier loading.

With a clear understanding of the design process, and the physical laws of nature that constrain it, it is easier to understand how the various approaches to process intensification have evolved into the options we have today. It also provides clues as to what direction technology development should be taking in the future. The three prevalent strategies to process intensification that are seen in the industry today are:

1. Improve settling characteristics,
2. Replace the clarifier with membrane filtration, and
3. Introduce a media supported biofilm that, because it is immobilized in the bioreactor, does not impact solids loading rates (SLR) to the clarifier.

While each of these strategies can legitimately claim to provide process intensification, they can lead to very different outcomes in terms of equipment requirements, level of mechanization and automation required, operational reliability, energy requirements and effluent quality. Or in the terms used during technology screening and selection: the capital cost, operating cost, reliability, and energy requirements/carbon footprint of the available strategies are not the same.

The following sections provide a brief overview of Strategies 1 and 2 for the purpose of better highlighting the features of Strategy 3, which is the subject of this book.

2.1.1 MEMBRANE BIOREACTORS

The intensification claim of membrane bioreactors (MBR) is that they enable operating at a mixed liquor suspended solids, MLSS, concentration of 8,000 to 12,000 mg/L, as opposed to the more typical 3,000 mg/L for conventional activated sludge. Above 8,000 to 12,000 mg/L, it becomes difficult to sustain good rates of permeation, or flux, through the membranes at reasonable transmembrane pressures, *TMP*.

The flux is the flow-rate per unit area of membrane and is commonly expressed in units of $L/m^2/h$ (LMH). The TMP is the pressure drop across the membrane and can be based on either a suction pressure from the permeate side, or a positive pressure from the mixed liquor side. The term "permeability" is used to describe the ratio of flux to TMP and is commonly presented with units of LMH/kPa or LMH/psi.

Flux

Typical rates of flux for MBR range from 11 to 30 LMH or 0.26 to 0.72 m/d. In comparison, secondary clarifiers are typically designed for surface overflow rates of around 40 m/d. Assuming a mixed liquor concentration of 10,000 mg/L, an equivalent SLR to membranes would range from 2.6 to 7.2 $kg/m^2/d$. In comparison, a design SLR of 170 $kg/m^2/d$ would be typical for clarifiers in a nitrifying activated sludge process.

So MBR's operate at an effective hydraulic loading rate that is 50 to 150 times lower than secondary clarifiers, and an effective solids loading rate that is 25 to 65 times lower. The conclusion is that membrane filtration cannot match the area specific loading rates of gravity based sedimentation. It should be noted, however, that MBR's remove much more particulate material than secondary clarifiers due to their small pore sizes, on the order of 0.04 μm. And so one of the arguments for the intensification benefits of MBR can be that it provides a level of TSS removal that is equivalent to secondary clarification plus tertiary filtration. That is to say, MBR's combine secondary and tertiary treatment into a single step.

Comparing Specific Surface Areas

The specific surface area of an MBR tank depends on the packing density of the membranes, and whether they are arranged as bundles of hollow fibers or flat sheets. A range of 150 to 300 m^2 per m^3 of tank volume is available from the various MBR equipment suppliers. In comparison, a circular secondary clarifier with a depth of four meters has a specific surface area of 0.25 m^2/m^3.

So a typical MBR tank offers an effective specific surface area of 600 to 1,200 times greater than a secondary clarifier. Herein lies the fundamental advantage of MBR's: the higher specific surface area achievable in a membrane tank more than offsets the lower loading rates per unit surface area, *i.e.* the high packing density overcomes the low flux.

Note that it is possible to increase the specific surface area of a clarifier through the use of lamella. However, this does not improve performance in clarifiers when zone settling is rate limiting, which is the case in most secondary clarifiers.

2.1.2 IMPROVING SETTLING PROPERTIES

Particles reach their terminal settling velocity when gravity forces F_g and drag forces F_d reach their equilibrium:

$$F_g - F_d = 0 \tag{2.1}$$

where:

$$F_g \propto \text{mass}$$
$$F_d \propto \text{size}$$

This relationship indicates that settling properties can be improved by maximizing particle mass (gravity forces) while minimizing particle size (drag forces). In fact, settling of mixed liquor in secondary clarifiers is not so straightforward. Mixed liquor flocs do not move as discrete particles, but rather as part of a sludge blanket. Referred to as zone settling, or hindered settling, the movement of the sludge blanket is slower than what would be expected from discrete settling of individual floc particles. Nevertheless, the relationships between gravity forces, drag forces and terminal velocity provide useful guidance to understand strategies for improving settling properties, and secondary clarifier capacity.

The following paragraphs describe technologies for improving mixed liquor settling properties by addition of ballasting agents, biological selection for granular sludge and mixed liquor degassing. It will be shown how each of these technologies is a strategy aimed at either increasing F_g or minimizing F_d.

Mixed Liquor Ballasting

Adding fine ballast that embeds itself into the mixed liquor flocs can dramatically increase the settling characteristics and debottleneck the secondary clarifier. The principle behind this strategy is that the ballast material has a density greater than that of biomass, which is only slightly more dense than water. Fine sand and magnetite are two ballast materials that meet this criteria with specific gravities of 2.65 and 5.2, respectively. For a given floc size, we can see how embedding ballast in the flocs can increasing gravity forces F_g by increasing the floc density ρ:

$$F_g = mg = \rho 4/3\pi R^3 \quad (2.2)$$

where:

ρ is the floc density
R is the floc radius assuming shape is a sphere

The challenge, however, is how to recover the ballast from the waste activated sludge? This is done in the BioMag process, marketed by Evoqua Water Technologies LLC, by acting on the magnetic properties of the magnetite to recover up to 95% of the ballasting material. The ability to load the secondary clarifier at two to three times higher MLSS is reported with this technology.

Biological Selection for Granulating Sludge

It has long been known that inclusion of unaerated "selector" zones, plug-flow reactors, and hydraulic "surface wasting" are process design features that can lead to improved mixed liquor settling characteristics [6]. More recently, processes are being developed which seek to more aggressively act on "the biology" and produce mixed liquor that looks and behaves more like granules than flocs. One definition for granular activated sludge is that it be of microbial origin, that does not coagulate under reduced hydrodynamic shear and [7]:

- settles significantly faster than activated sludge, with minimal difference between SVI_{10} and SVI_{30}, and
- has a minimum particle size of 200 μm.

Based on this definition, the obvious explanation for the improved settling velocity of granular sludge is size. Granular sludge with a particle size of 200 μm is subject to much lower drag forces, relative to its mass, than activated sludge particles with a particle size typically less than 50 μm.

The relationship between drag forces, mixed liquor particle size and the terminal settling velocity is described according to Stoke's Law presented below. Even though the force of drag F_d increases in proportion to particle R, as shown in Equation 2.2, the gravity forces increase in proportion to R^3. Therefore larger particles settle much faster.

$$F_d = 6\pi\eta R v \tag{2.3}$$

where:

F_d is the frictional force or drag acting on the particle
η is the dynamic viscosity
R is the radius assuming the particle is a sphere
v is the settling velocity

It should be remembered, however, that performance of secondary clarifiers is governed by zone settling velocities, and compressibility of the sludge at the bottom of the sludge blanket. The enhanced settling characteristics of granules, as compared to flocs, is helpful but may not directly translate into increased limits on allowable secondary clarifier loading rates. State-point analysis is the standard tool for relating settling properties of sludge to clarifier capacity [6].

If one accepts the premise that granular sludge enables process intensification, the question is how to achieve this? Royal HaskoningDHV has pioneered the application of aerobic granular sludge technology through their proprietary Nereda process. Selection for granular activated sludge in the Nereda process seems to rely on a combination of the following factors:

- batch feeding to generate high substrate gradients at the bottom of the reactor creating an enhanced "selector" effect,
- imposition of high shear conditions, and
- selective wasting of poor settling material from the bioreactor.

It is also known that influent wastewater strength and availability of volatile fatty acids play an important role in promoting sludge granulation. It is likely that sludge granulation may not be feasible in all situations. In addition, there are concerns in purely granular reactors about potentially losing the "sweeping" action of flocs, which is important for minimizing suspended solids concentrations in the secondary clarifier effluent.

Mixed Liquor Degassing

Mixed liquor degassing essentially applies the same strategy as mixed liquor ballasting, which is to increase the density of the flocs. The only difference is that, whereas ballasting aims to add particles to the floc that are denser than biomass, mixed liquor degassing aims to take away microbubbles that are considerably less dense.

Microbubbles can form in the mixed liquor flocs and negatively impact settling characteristics in the secondary clarifier. These microbubbles can result from denitrification in post-anoxic zones, *i.e.* nitrogen gas. It is also possible to form microbubbles in purely aerobic processes. This is known to occur when mixed liquor flows from a deep aeration tank to a shallow secondary clarifier. The release in pressure as mixed liquor flows from a deep tank to a shallow clarifier leads to a supersaturation condition. Excess carbon dioxide and nitrogen gas is released in microbubbles that can adhere to the activated sludge flocs.

The traditional way for managing the effect of microbubbles is to include a small zone to air sparge the mixed liquor prior to flow into the secondary clarifier. The intent is to use the energy of large bubbles to knock the little bubbles out of the flocs. Another approach being marketed under the tradename Biogradex subjects the mixed liquor to vacuum conditions to remove microbubbles by "vacuum degasification". Capacity enhancements using this approach are reportedly greater than 30%. Incidentally, 30% is typically the threshold savings requirement for overcoming the risk of adopting new technologies.

2.1.3 IMMOBILIZING SLUDGE IN THE BIOREACTOR

There is a long history, dating back to the 1930s, of introducing media-supported biofilms into activated sludge bioreactors. Whereas the previous intensification strategies focussed on improving settling properties in the clarifier, or replacing the clarifier altogether, hybrid processes immobilize a portion of the sludge in a biofilm so that it never reaches the clarifier at all. The result is essentially a two-sludge system: one that exists in the biofilm and the second in the mixed liquor.

With a portion of the biomass immobilized in the biofilm, there is a potential to double or even triple the total sludge inventory in the bioreactor while maintaining the same solids loading rate to the clarifier. For illustrative purposes, a case where the biofilm sludge could effectively double the total sludge bioreactor sludge mass is worked out below.

Equivalent Sludge Mass

Assuming an MLSS of 3,000 g/m^3 and volume of V of 1,000 m^3, we get the following mixed liquor sludge inventory, M_{MLSS}:

$$M_{MLSS} = MLSS \times V = 3,000\,g/m^3 \times \frac{1\,kg}{1000\,g} \times 1000\,m^3 = 3,000\,kg$$

And assuming a media specific surface area, SSA, of 500 m^2/m^3 and a fill fraction of media in the bioreactor of 50%, we get a total surface area, A, as follows:

$$A = SSA \times F_{Fill} \times V = 1000\,m^3 \times 0.5\,m^3/m^3 \times 500\,m^2/m^3 = 250,000\,m^2$$

Now assuming a moderate biofilm coverage, B_{Solids}, on the media of 10 g/m^2, we get a biofilm sludge inventory of:

$$M_{Biofilm} = A \times B_{Solids} = 250,000\,m^2 \times 10\,g/m^2 \times \frac{1\,kg}{1000\,g} = 2,500\,kg$$

So in the above equations the total bioreactor sludge inventory for the hybrid process ($M_{MLSS} + M_{Biofilm} = 3,000\,kg + 2,500\,kg = 5,500\,kg$) is roughly double what it would have been in a conventional process. It's not hard to see how the contribution of the biofilm could further increase if we increase the assumed biofilm coverage B_{Solids} from 10 to 20 or 30 g/m^2.

Diffusion Limitations

With respect to increasing sludge inventory in the bioreactor, hybrid processes achieve similar levels of intensification as MBR. However, it is well known that treatment performance in a biofilm is subject to diffusion limitations and thus is not equal to treatment performance in mixed liquor. That is to say, that 1 kg of biofilm sludge does not provide the same treatment performance as 1 kg of mixed liquor sludge.

Quantifying Benefits

Quantifying the benefits of biofilms in hybrid systems can be challenging, particularly where the behavior of the biofilm and the mixed liquor are interdependent. Tools available for addressing this challenge discussed in this book include:

- Bench scale testing to isolate the activity of media supported biofilm and mixed liquor.
- Microbial techniques to quantify the relative heterotrophic and nitrifying composition of the biofilm.

- *In situ* measurement of oxygen transfer rates.
- Use of design curves to establish relationship between biofilm and mixed liquor nitrification performance.
- Simulation of steady-state and dynamic performance using process modeling software.

Each of these tools provides a unique perspective for understanding the performance and behavior of hybrid systems. However, tying this information together into a coherent story that communicates the bottom line benefits for process intensification, one that is both clear and accurate, is not a simple task. And sharing this story with project stakeholders and decision makers who often have no training in process engineering, i.e. a "non-technical" audience, can be very challenging indeed. One of the purposes of this book is to better equip engineers to communicate the bottom line benefits of hybrid systems.

2.2 HYBRID TECHNOLOGIES

The term IFAS has become the most commonly used term for hybrid processes in the industry. It will be used in this book to generically refer to hybrid processes including 1. Fixed-media/Activated Sludge (Fixed-media/AS), 2. Moving Bed Biofilm Reactors/Activated Sludge (MBBR/AS), and 3. Membrane Aerated Biofilm Reactors/Activated Sludge (MABR/AS). These three categories adequately capture the main hybrid processes that we commonly encounter in practice today.

The IFAS technologies we see in practice today are the result of lessons learned from experimentation over the decades with different reactor designs, as well as types and configurations of media. Asbestos sheets were used in the 1930s and 1940s for the contact aeration process, wooden planks in conventional activated sludge tanks in the 1970s, woven textile media and mobile media in the 1990s, and now membrane aerated "MABR" media in the 2010s. This evolution has in large part been a response to challenges encountered with respect to biofilm thickness control, hydraulics and overcoming biofilm diffusion resistance, which has led us to the IFAS we see today. The following sections provide an overview of Fixed-media/AS, MBBR/AS and MABR/AS.

2.2.1 FIXED MEDIA/AS

While earlier approaches to using fixed media included asbestos sheets, wooden boards and submerged rotating biological contactors, current practice is more likely to use woven rope, textiles fibers or mesh-type media arranged in a frame. Examples of proprietary rope-like material are Ringlace and BioMatrix. Examples of the mesh-type are AccuWeb and BioWeb. Early implementations of these types of media were plagued by problems of poor biomass thickness control and worm overgrowth. This unfortunately led to the technologies being removed in some cases.

Biofilm Thickness Control

Providing adequate scouring energy to manage biofilm thickness is arguably the most important issue for a successful Fixed-media/AS process. Too low of a biofilm thickness would be limiting treatment performance; after all, without enough organisms in the biofilm there is no treatment enhancement, but this is rarely the problem in practice. Too high of a biofilm thickness, however, is a frequently encountered problem. The main reason that thick biofilms are a problem is because it can cause bridging between media elements. Bridging between media elements dramatically reduces the effective surface area, and for a surface diffusion limited process, the result is dramatically reduced treatment performance. It is axiomatic that treatment performance in biofilms is limited by surface diffusion.

The end result of poor biofilm thickness control can be that media frames turn into a solid mass of sludge in which all of the interior surface area of the media has been lost. This is illustrated in Figure 2.2 where a specific surface area of 315 m^2/m^3, as quoted by one Fixed-media vendor, is achieved when the biofilm is thin and individual media elements, in this case textile cords, behave independently of one another. What this means is that the surface area of each of the individual media elements is equally exposed to the bulk liquid ammonia and dissolved oxygen concentrations.

But when the interior surface is lost due to excessive biofilm thickness or "sludging", the effective surface area is only the outside six faces of the media frame, so 6 m^2/m^3. Dropping the specific surface area from 315 to 6 m^2/m^3 effectively reduces the process performance to nil. Worse yet, the displaced liquid volume of Fixed-media can increase from something less than 10% to 100%. In this case, not only is the biofilm not contributing to overall hybrid process performance, but the displaced bioreactor volume diminishes the treatment performance of the mixed liquor as well.

Biofilm Sludge Management

In addition to loss of treatment performance, high biofilm thickness can lead to serious operational problems related to managing the biofilm sludge. When biofilm thickness is excessive, the media frame may not be able to support the weight when the tank is drained and the biofilm is no longer buoyant. Conditions inside the biofilm sludge are likely to turn anaerobic and this can lead to odor problems. These types of problems explain resistance to Fixed-media/AS as an intensification alternative in many activated sludge upgrade projects. Fixed-Media/AS equipment vendors have modified their designs over the years, for example by improving biofilm scouring strategies, to mitigate these problems.

Red Worms

Control of red worm has been a historical problem for Fixed-Media/AS. As red worms are obligate aerobes, their blooms seem to stem from operation at high dissolved oxygen concentrations, which may occur during low loading periods. Also, higher biofilm thicknesses seem to promote their growth.

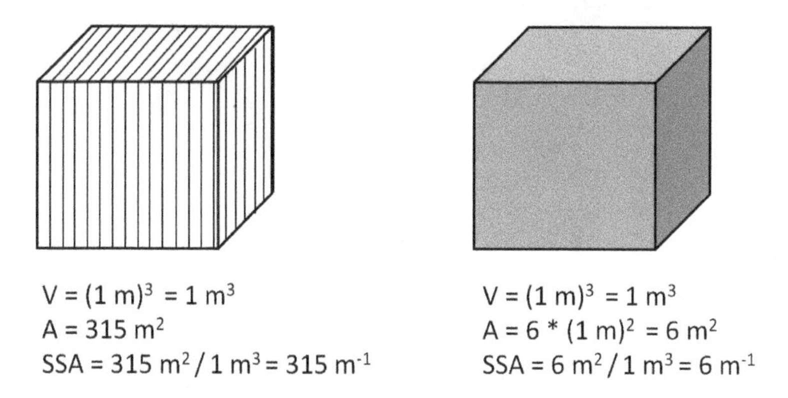

Figure 2.2 Comparison of specific surface area of Fixed-media unit with and without adequate biofilm thickness control

The presence of red worms at some level may be acceptable and even beneficial since they contribute to lower sludge yields. However, excessive blooms can be undesirable as a nuisance problem, and it is reported that their excessive grazing of biofilms can lead to reduced treatment performance. Overall, the conditions that lead to excessive red worm growth are not fully understood and the topic receives little attention from researchers.

Strategies for controlling red worm growth mainly relate to limiting dissolved oxygen concentrations in the bulk liquid. In addition, periodically turning off the aeration to the mixed liquor for a period of several hours is reported to be effective.

MABR

MABR is a recently commercialized type of Fixed-media/AS in which aeration is provided inside the media lumen. Many of the considerations identified above with respect to biofilm thickness control, biofilm sludge management and red worms also apply to MABR. However, there are a number of unique features of MABR that make them very different from traditional Fixed-media/AS. These will be discussed in Section 2.2.3.

2.2.2 MBBR/AS IFAS

Moving bed biofilm reactors, MBBR, using plastic biofilm carriers were originally developed as "flow through" processes, that is, processes without any clarifier return activated sludge. Sometimes referred to as "pure biofilm" processes, essentially all of the treatment must take place in the biofilm because there is no mixed liquor. MBBR media have since been used in hybrid IFAS processes where treatment occurs both

Figure 2.3 Meteor Media, courtesy SUEZ Water Technologies & Solutions

Figure 2.4 Bioreactor tank with Meteor Media, media retention sieves and overflow weir, courtesy SUEZ Water Technologies & Solutions

in the biofilm and the mixed liquor. Typically these are retrofits to existing activated sludge processes where the role of the biofilm is to supplement the treatment capacity of the mixed liquor. In this book we will refer to IFAS processes that utilize mobile media as "MBBR/AS". As compared to Fixed-media/AS, MBBR/AS has the advantage that, due to media mobility, biofilms are subject to greater shear forces. This makes it easier to control biofilm thickness.

The pioneering company for MBBR and MBBR/AS processes was Kaldnes Miljoteknolog using a wagon wheel style plastic "Kaldnes" media of about 10 mm in height. There are now numerous vendors offering a range of shapes and sizes. As an example, the Meteor media used by SUEZ Water Technologies & Solutions is presented in Figure 2.3. The amount of competition in the marketplace for MBBR media and equipment reflects the success the technology has had.

MBBR/AS is one of the standard technologies for activated sludge intensification and is used around the world for both municipal and industrial applications. In many cases, however, the very fact that the media is "mobile" has been an obstacle to its adoption. The need to include retention sieves in the bioreactors to hold the media in place can create a bottleneck in the plant hydraulic gradeline and increases the risk of flooding in the bioreactor. Though flooding events are rare, and can be managed using overflow weirs and channels, publicized events of media loss into the receiving water environment has created a negative stigma around this technology in some quarters [37].

Physical requirements that must be carefully considered in the design and operation of MBBR/AS include:

Sieves

Media retention sieves must be located at the effluent end of the bioreactor as well as all potential overflow locations. As shown in Figure 2.4, the retention sieves may be cylindrical and protrude into the tank so as to increase their surface area. Typical design is for a hydraulic loading rate of 50 to 60 m/h for the peak flow condition.

The size of holes in the sieve should be smaller than the media size but large enough to allow trash and other debris to pass. Use of perforated plate or wedge wire sieves is typical.

Foam Removal

Foaming bacteria can grow and be trapped in activated sludge bioreactors where the effluent discharge is below the surface. Excessive foaming is a nuisance for operators and its control may require installation of spray nozzles with chlorine dosing. Good design practice for conventional activated sludge avoids foam trapping by allowing for mixed liquor to overflow between bioreactor zones. This design practice is not compatible with MBBR/AS retrofits. Because media is retained within MBBR bioreactors using submerged media retention sieves, accumulation of foam is a common issue. The use of higher airflow rates in MBBR zones further contributes to the problem.

Mixing and Scour

Adequate mixing must be provided to ensure that free floating MBBR media remains uniformly distributed in the bioreactor and do not accumulate at the downstream end of the tank. In some cases, it may be necessary to include an airlift pump to ensure even distribution of media between upstream and downstream ends of the bioreactor. In addition, it is common practice to use an airknife below the media retention sieves to prevent clogging from media and trash. Clogging of the sieves would lead to bioreactor tank flooding.

Coarse bubble aeration grids are typically arranged to create a rolling pattern in the bioreactor. The advantage of coarse bubble aeration is that the larger bubbles provide greater mixing and scouring energy. The latter is critical for maintaining good biofilm thickness control. In addition, a stainless steel coarse bubble diffuser is strong enough to resist the weight of the media when the tank is drained. With a fine bubble grid, media would need to be removed from the tank prior to draining the tank.

Aeration

Despite the use of stainless steel coarse bubble diffusers, MBBR bioreactor aeration is commonly referred to as "medium bubble" because the oxygen transfer efficiency (OTE) observed is better than would be expected for conventional coarse bubble aeration. This is attributable to the retention of bubbles on media which increases their retention time in the bioreactor water column, thereby improving the transfer efficiency. Oxygen transfer efficiency in MBBR bioreactors is still lower than conventional activated sludge, however, due to the need to operate at higher bulk liquid dissolved oxygen concentrations, typically 3 to 4 mg O_2/L.

Headworks Screening Requirements

For raw sewage, headworks screening requirements for MBBR/AS are typically reported as 6 mm or less. Equipment vendors have an incentive to allow for 6 mm because this is the standard to which most conventional activated sludge plants have been designed. However, it is well known that MBBR/AS processes report much fewer operational problems with finer screens. This is true of plant equipment in general including pumps, analytical sensors and biosolids handling processes.

Media Selection

Media is typically selected to provide a high specific surface area. In this respect, smaller media have the advantage. Just consider the surface area to volume ratio of a 1 mm radius sphere compared to a 10 mm radius sphere: $\frac{4\pi R^2}{4/3\pi R^3}$, where R is the radius. Media also needs to be large enough to be retained in media retention sieves, however. And the media retention size hole openings need to be large enough to allow trash of 6 mm, or even greater, to flow through. It is not uncommon to find trash particles in the mixed liquor of greater size than might be dictated by headworks screening. This can be attributable to poor capture or bypass occurrences in the headworks screens, trash that the wind blows into the bioreactors and clarifiers, or even "roping" which converts smaller into larger size trash particles.

In practice, media dimensions used for MBBR/AS are around 20 to 30 mm in diameter and have a specific surface area around 500 to 800 m^2/m^3. An additional consideration for plastic media selection is that it include additives to resist UV degradation. In the absence of this, the media will be much more prone to breakage, which could result in media fragments being carried with the plant effluent into the receiving water environment.

MBBR Nitrification Rates

MBBR/AS processes are typically designed based on a specific nitrification rate (NR) on the order of 0.5 g $N/m^2/d$ and only for conditions where BOD loading is low. For pure MBBR processes, this means MBBR as a tertiary stage process, or designing a 2-stage MBBR process in which the first stage provides BOD removal and the second stage provides nitrification. In hybrid MBBR/AS processes, the implication is that installation of media at the upstream end of the bioreactor does not enhance nitrification capacity and is generally avoided.

The possibility of achieving significantly higher nitrification rates, as high as 2.5 g $N/m^2/d$, has been demonstrated by Hem *et al.* for conditions where BOD loading is absent and dissolved oxygen is very high, 10 mg O_2/L [11]. However, this is very difficult to achieve in practice and even for a tertiary treatment application, where BOD loading should have effectively been absent, Hem *et al.* only reported nitrification rates of 0.7 to 1.0 g $N/m^2/d$ at bulk dissolved oxygen concentrations of 5 mg O_2/L. Similarly, Houweling *et al.* reported nitrification rates in a range of 0.1 to 0.8 g $N/m^2/d$ for a second stage MBBR treating effluent from a municipal lagoon at dissolved oxygen concentrations as high as 8 mg O_2/L [15]. So despite the potential for achieving much higher nitrification rates, designing MBBR/AS processes based on a nitrification rate of approximately 0.5 g $N/m^2/d$ is appropriate [9].

Oxygen-limited Nitrification

As stated by Ødegaard: "As long as ammonia concentration is above 1-2 mg NH_4^+/L ... the nitrification rate will not be limited by ammonium, but by oxygen concentration" [36]. The threshold for transition between ammonia-limited and oxygen-limited conditions varies based on bulk liquid dissolved oxygen concentrations which, according to Hem *et al.*, occurs at a ratio of about 3 mg O_2/mg NH_4^+-N [11]. So achieving ammonia-limited behavior in a range of 0 to 2 mg NH_4^+/L would require operation at a dissolved oxygen concentration of 6 mg O_2/L. This is about the limit of what could reasonably be achieved in practice and, for hybrid processes, bulk liquid concentrations of 3 to 4 mg O_2/L are far more typical. It should be remembered that these thresholds were derived for pure MBBR processes, *i.e.* with no mixed liquor, and for reactors without any BOD loading. But the lessons learned for hybrid systems are the same: nitrification rates in MBBR biofilms are primarily oxygen-limited and show no response to bulk liquid ammonia until low concentrations, below 1 to 2 mg NH_4^+/L, are reached.

Whether nitrification rates are ammonia-limited, oxygen-limited, or even biomass limited, has important consequences for hybrid process design, notably the ability to manage peak load events. Ammonia-limited nitrification is preferable in this regard because it means that nitrification rates will increase to match peak loading events: as ammonia concentrations in the bulk liquid increase, so will the nitrification rate. Ammonia-limited nitrification in a range of 0 to 1 or 2 mg NH_4^+/L for MBBR biofilms is higher than what would be expected for mixed liquor flocs, which might be expected to be ammonia-limited in a range of approximately 0 to 0.5 mg NH_4^+/L. The benefits for process design are discussed further in Section 2.6, "Implications for Ammonia- and Oxygen-Limited Biofilms", as well as in Chapter 6.

2.2.3 MABR/AS IFAS

The potential of combining oxygen transfer through membranes with biofilm treatment was recognized as early as the 1960s and 1970s [28, 17]. This remained a topic of interest for academic and industrial researchers throughout the 1980s, 1990s and 2000s. However, only in the last five years have commercial products been brought to market under the commonly used name "Membrane Aerated Biofilm Reactors" or MABR. Using both hollow-fiber and spiral-wound membrane approaches, commercial products include "ZeeLung" marketed by SUEZ Water Technologies & Solutions, "OxyFas" marketed by Oxymem Ltd., and "Aspiral" marketed by Fluence Corp.

Comparing MABR to MBBR

MABR and MBBR biofilms are presented schematically in Figure 2.5 to highlight the following differences between the two technologies.

- In the MBBR biofilm (A), both ammonia and oxygen co-diffuse into the biofilm from the bulk liquid. In the MABR biofilm (B), oxygen counter-diffuses into the base of the biofilm from the media lumen.
- Because oxygen diffuses into the biofilm from the media, there is no requirement for bulk liquid dissolved oxygen in the MABR biofilm (B). Aerobic conditions, however, are not incompatible with MABR operation.
- In the absence of dissolved oxygen in the bulk liquid of (B), nitrate generated in the biofilm can be denitrified in the bulk liquid mixed liquor.
- MABR media (B) is smaller in diameter than MBBR media (A), respectively, typically on the order of 1 mm compared to 20 mm. This is true of the hollow-fiber support media pictured in (B) and would not be true for a spiral-wound support media.
- To minimize resistance to oxygen diffusion into the biofilm, the membrane wall thicknesses of MABR media (B) are very thin, on the order of 0.1 mm.

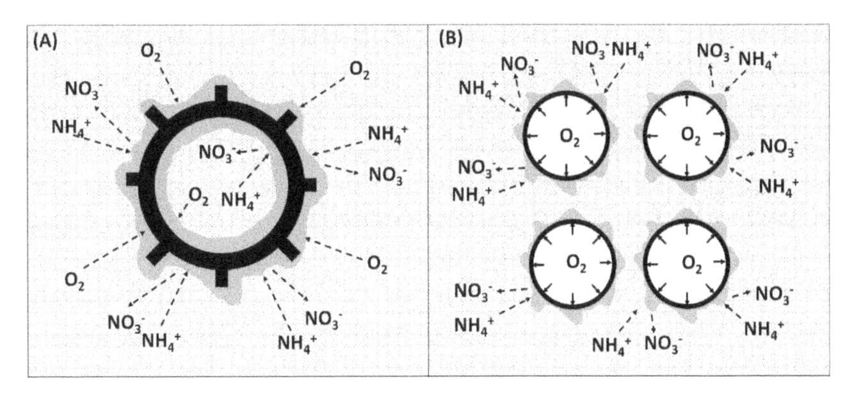

Figure 2.5 Schematic representations of (A) MBBR-media supported biofilm and (B) MABR-media supported biofilm

- Oxygen availability in the MBBR biofilm (A) is proportional to the dissolved oxygen concentration in the bulk liquid and the respiration rate of the biofilm organisms. Bulk dissolved oxygen concentrations in MBBR/AS processes are typically 3 to 4 mg O_2/L.
- Oxygen availability in the MABR biofilm (B) is proportional to the partial pressure of oxygen in the media lumen and the respiration rate of the biofilm organisms. The partial pressure of oxygen is typically on the order of 20 to 30 kPa but can be increased by operating at higher lumen air pressures, or by supplying air enriched in oxygen.

The ability to tune oxygen availability in an MABR biofilm through adjustments to lumen air pressure is an attractive feature of this technology. It provides a more effective, and energy-efficient, means for avoiding oxygen-limiting conditions than is possible in MBBR biofilms which must rely on adjustments to bulk liquid dissolved oxygen concentrations.

Unique Properties of MABR Biofilms

The counter-diffusional properties of MABR biofilms, illustrated in Figure 2.5, have been studied extensively by researchers with respect to the stratification of heterotrophic and nitrifying bacteria, and the impact of BOD loading on nitrification rates [3, 20, 8]. It was shown by Downing and Nerenberg (2008) that nitrification in MABR biofilms is less sensitive to BOD loading than conventional, co-diffusional biofilms [8]. The explanation provided is that nitrifying organisms grow in the deep, aerobic portions of the biofilm where BOD concentrations are low, and heterotrophic organisms grow on the outer portions of the biofilm using nitrite and nitrate as the electron acceptor, *i.e.* denitrification. So counter-diffusional biofilms promote nitrification by supplying oxygen at the location where nitrifying organisms are most abundant. An additional factor that may be at play is the diffusivity of ammonia which, due to its low molecular weight, is higher than the diffusivity of BOD constituents [30]. So not only is BOD consumed on the outside of the biofilm through denitrification, it also diffuses into the biofilm more slowly than ammonia.

The unique opportunities afforded by MABR for IFAS upgrades are summarized as follows:

1. The oxygen transfer efficiency (OTE) in MABR's can be very high, as high as 100 %. MABR therefore provides a means to reduce the energy requirements and carbon footprint of hybrid processes.
2. Oxygen diffusion rates into the biofilm can be several times higher than in conventional systems, thereby boosting nitrification rates.
3. Less inhibitory effect of BOD on nitrification due to the unique properties of the counter-diffusional biofilm.
4. Potential to operate in unaerated zones where only the biofilm is aerobic and mixed liquor is anoxic or anaerobic.

MABR Nitrification and Oxygen Transfer Rates

Performance metrics are summarized in Table 2.1 for five demonstration scale MABR/AS installations [16]. Based on results from Table 2.1, design using biofilm nitrification rates of 1.5 to 2.4 g $N/m^2/d$ appears appropriate for MABR/AS. It should be noted, however, that bulk ammonia concentrations were primarily in the range of 10 to 20 mg N/L for these MABR/AS installations. Lower nitrification rates would be expected for MABR biofilms operated with bulk ammonia concentrations in the range of 0 to 10 mg N/L.

The results from Table 2.1 provide a dramatic contrast to nitrification rates presented in Section 2.2.2 for MBBR biofilms, which were on the order of 0.5 to 1 g $N/m^2/d$. Higher MBBR nitrification rates would certainly be possible if oxygen limited conditions in the biofilm could be addressed. Hem *et al.* indicated the possibility of 2.5 g $N/m^2/d$ at very high dissolved oxygen concentrations, and in the absence of BOD loading [11]. But achieving these conditions in hybrid MBBR/AS processes is not feasible: transferring oxygen into mixed liquor at concentrations above 3 to 4 mg O_2/L is simply cost prohibitive. It would require too much energy and aeration capacity. Adjustment to the lumen air pressure in MABR media offers a more powerful, and energy efficient, means for avoiding oxygen-limited conditions in biofilms. This explains the dramatic improvement in nitrification rate they afford.

Table 2.1
Summary of Oxygen Transfer Rate, Nitrification Rate and Oxygen Transfer Efficiency for MABR/AS Process [16]

Percentile	Nitrification Rate g $N/m^2/d$	Oxygen Transfer Efficiency %	Oxygen Transfer Rate g $O_2/m^2/d$
25th	1 to 2	25 to 34	6.8 to 10
50th	1.5 to 2.4	28 to 37	7.5 to 11.3
75th	1.7 to 2.8	30 to 40	8.7 to 12

MABR in Anoxic Zones

The ability to install MABR media in anoxic bioreactor zones provides a means to enhance both the nitrification and denitrification capacity of the zone. Compared to a conventional BNR process, where the purpose of the anoxic zone is denitrification only, an anoxic zone with MABR media can simultaneously nitrify in the biofilm and denitrify in the mixed liquor. Simultaneous nitrification/denitrification (SND) is highly valued as a means to maximize total nitrogen removal in the available bioreactor zone.

Many plants operate aeration strategies to achieve SND in conventional biological nutrient removal (BNR) plants. These typically involve maintaining dissolved oxygen in the range of 0.5 to 1 mg O_2/L so that both nitrification and denitrification are promoted in the flocs [18]. But neither reaction rate is optimal under these conditions. In addition, mixed liquor that is neither fully anoxic nor fully aerobic can lead to excessive growth of filamentous organisms, which in turn leads to bulking sludge. Bulking sludge, or even the risk of it, means poorer settling properties need to be accounted for in clarifier operations. The net result is that conventional aeration strategies to achieve SND may derate the capacity of the clarifiers.

SND is a valued operational strategy for achieving total nitrogen removal and strategies have been developed over time to avoid problems, such as bulking sludge, that may result. MABR provides a new strategy for achieving the benefits of SND without the nuisance problems of bulking sludge.

Operational Concerns

Because MABR media is mounted in frames, it shares a lot of the same design and operational concerns as Fixed-media/AS, notably how to control biofilm thickness and avoiding excessive growth of red worms. The potential to install MABR media in unaerated zones significantly lowers the risk of excessive growth of red worms, which are obligate aerobes.

Regarding controlling biofilm thickness, many of the lessons learned from Fixed-media/AS processes have been applied to MABR/AS. In particular the need to ensure adequate scour energy using aeration systems integral to the MABR product has been recognized. Indeed a lot of the design features from commercial equipment vendors are related to aeration systems, integral to the MABR cassettes or modules, aimed at providing effective biofilm scouring.

2.3 IFAS CONFIGURATIONS

Fixed-media, MBBR and MABR media can be used for upgrade of all kinds of activated sludge process configurations including sophisticated biological nutrient removal (BNR) configurations incorporating multiple unaerated zones. The appropriate location for media placement depends on what treatment process the biofilm is meant to enhance: BOD removal, nitrification or denitrification. The most common role for the biofilm nitrification.

BOD Removal

Effective removal of soluble BOD and, to a lesser extent particulate BOD, in a biofilm can be achieved when located at the upstream end of the bioreactor. The value to the overall process is limited, however, as BOD removal in an activated sludge process tends to be very good, even at SRTs as low as two or three days. Below this threshold, mixed liquor flocculation degrades and BOD and TSS removal are poor. A hybrid process would not be expected to help mitigate problems with mixed liquor flocculation.

Nitrification

To achieve a nitrifying biofilm, media must be placed in a location where there is sufficient ammonia, and soluble BOD loading is low enough that heterotrophic organisms will not outcompete nitrifiers for the available dissolved oxygen. The downstream end of the bioreactor, where ammonia is too low to sustain growth of a healthy nitrifying biofilm, is therefore not appropriate. Media placed at the upstream end of the bioreactor is also unlikely to support the growth of a nitrifying biofilm. The best location for media placement, where BOD loading is not too high and ammonia concentrations are not too low, is therefore in the middle of the bioreactor. This is illustrated in Figure 2.6 for an MBBR/AS process.

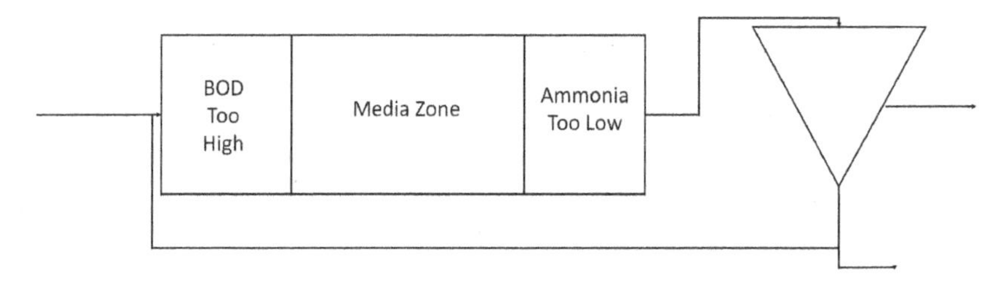

Figure 2.6 Illustration of bioreactor zones appropriate for a nitrifying biofilm in an MBBR/AS IFAS configuration

Because nitrification in MABR biofilms is less sensitive to BOD loading than it is in MBBR biofilms, it is possible to place media further upstream in the process. This is illustrated in Figure 2.7 which shows the size of the zone where BOD loading is considered "too high" as smaller than in Figure 2.6.

It should be noted that experience with MABR/AS at full-scale is still very recent. A wide range of conditions can influence how BOD loading impacts nitrification, for example: oxygen partial pressure in the media lumen, dissolved oxygen and nitrate in the mixed liquor, biofilm thickness, and influent wastewater composition, to name a few. It is expected that much will be learned in the coming years on the influence of BOD loading on nitrification in MABR biofilms.

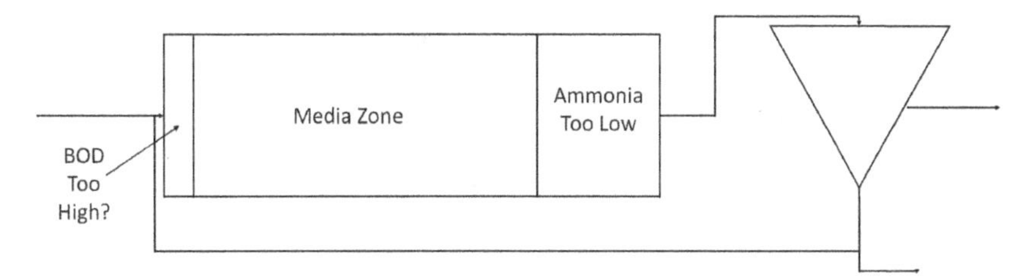

Figure 2.7 Illustration of bioreactor zones appropriate for a nitrifying biofilm in an MABR/AS IFAS configuration

Denitrification

Because denitrification in a pre- or post-anoxic zone can be biomass limited, increasing the sludge inventory using unaerated biofilms can provide enhancement to denitrification capacity. Denitrifying biofilms are most often used in post-anoxic zones where there is external carbon addition.

Enhanced Biological Phosphorus Removal

Enhanced biological phosphorus removal (EBPR) relies on subjecting heterotrophic organisms to alternating anaerobic and aerobic conditions which promotes the growth of poly-phosphate accumulating organisms (PAOs). Media supported biofilms cannot satisfy these conditions on their own, but they can add value to an EBPR process by contributing to the nitrification or denitrification step.

2.4 PROCESS DESIGN OF HYBRID PROCESSES (IFAS)

A typical IFAS application involves adding media to a conventional activated sludge process to enable nitrification where the SRT is otherwise considered to be too short. Example cases include:

- A new effluent permit has been issued which requires ammonia removal and the current activated sludge process was designed only to provide BOD and TSS removal,
- A new effluent permit has been issued with a total nitrogen (TN) limit which requires converting a portion of the bioreactor into anoxic zones to promote denitrification. The resulting loss of aerobic volume fraction means the aerobic SRT is no longer sufficient for reliable nitrification.
- A new effluent permit has been issued with a total phosphorus (TP) limit and the preferred option for meeting this limit is enhanced biological phosphorus removal. A portion of the bioreactor needs to be converted to an anaerobic zone and the resulting loss of aerobic volume fraction means the aerobic SRT is no longer sufficient for reliable nitrification.
- Plant flows and loads are increasing and the resulting increase in sludge production makes it impossible to maintain the SRT required for reliable nitrification.

Of course, construction of new bioreactors and clarifiers could be the solution to each of the problems identified above. However, this is not feasible, or not practical, in many cases due to financial or footprint constraints.

Just like for the activated sludge process, there are a variety of approaches used to design IFAS processes. The Water Environment Federation Manual of Practice No. 35 "Biofilm Reactors" (MOP 35) presents methods based on 1. empirical relationships, 2. pilot studies, 3. biofilm rate models and 4. using simulation software [9]. These methods are discussed below.

2.4.1 EMPIRICAL METHODS

Equivalent SRT

The equivalent SRT is an intuitive metric in which the mass of biofilm is seen to contribute to the total sludge inventory of the bioreactor when calculating the SRT:

$$SRT_{Equivalent} = \frac{M_{MLSS} + M_{Biofilm}}{Q_{WAS}TSS_{WAS}}$$

This method is used by some suppliers of sponge-media systems but is not recommended for use in Fixed-media/AS or MBBR/AS. The concept of equivalent SRT may be appropriate in sponge media systems due to the regular application of mechanical force to squeeze excess biomass out of the media into the bulk liquid. A higher rate of exchange between solids grown in the biofilm and the mixed liquor would justify treating the two sludges in an equivalent manner when calculating SRT.

Quantity of Media

An alternative empirical approach may be to use a "Quantity of Media" approach where each unit of media is assumed to provide "X" kg/d of treatment capacity, and "X" is derived from past experience. Such an approach may be useful for answering basic questions like "How much media would it take to solve my nitrification problem?" but should not be used for more than basic preliminary sizing.

2.4.2 PILOT STUDIES

Pilot studies are often used to simulate the proposed full-scale application at pilot-scale. They provide the following opportunities:

- Plant staff can become familiar with the technology.
- Process design can be validated and full-scale performance targets can be tied to the pilot study results.
- Pilot studies provide data that can be used for calibration of a process model. The process model can then serve as the basis of design for the full-scale plant.
- Pilot studies provide an opportunity to identify technical problems, and lessons learned can be applied to the full-scale design.

It should be noted that performance at pilot-scale cannot fully simulate what will occur at full-scale due to dimensional limitations such as water depth and basin geometry. In addition, performance at pilot-scale is often worse than at full-scale due to challenges such as: 1) less attention from operations staff, 2) smaller equipment that often is not purposely selected for the given piloting conditions, 3) difficulties in operating clarifiers at pilot scale, *etc.* As a consequence, pilot results should be interpreted with caution and, in most cases, should be complemented by some form of process modeling.

2.4.3 BIOFILM RATE MODEL

MOP 35 presents a set of rules-of-thumb that link reduction in mixed liquor SRT in a hybrid process to the fraction of the influent ammonia and COD load that are removed in the biofilm. This approach is based on observations in full-scale plants, pilot studies and in calibrated process models and can be summarized as follows [9]:

1. Identify nitrification rate based on temperature and organic loading.
2. Select a fraction of the influent ammonia load to be removed in the biofilm $F_{Nit,B}$ based on the mixed liquor SRT and temperature and assuming a target effluent ammonia concentration of 1 mg N/L. The lower the SRT the higher the $F_{Nit,B}$.
3. Select a nitrification rate assuming 25%, 50% and 75% of the maximum rate identified in Step 1 will be achieved in the first, second and third portions of the aerobic zone.
4. Calculate the media requirement from Steps 2 and 3.

Note that Step 3 recommends applying significantly lower nitrification rates at the upstream portion of the bioreactor due to the negative impacts of BOD loading on oxygen availability in the biofilm.

The required fraction of ammonia removal in the biofilm, $F_{Nit,B}$, from Step 2 is provided as follows: at an SRT of 2 d, required $F_{Nit,B}$ is 80%; at an SRT of 4 d, required $F_{Nit,B}$ is 50%; at an SRT of 8 d, required $F_{Nit,B}$ is 20%. These values are for a temperature of 15°C and guidance is provided on how they can be adjusted with temperature but not, as would be expected, the effluent ammonia objective. In Chapter 4, design equations are presented that extend these relationships to any combination of temperature, and effluent ammonia objective.

2.4.4 MASS BALANCING BIOFILM AND MIXED LIQUOR

Both MOP 35 and Metcalf & Eddy present a semi-empirical process design approach for IFAS processes based on work by Sen & Randall [9, 1, 29]. The basis of this approach is performing two separate mass balances: one to calculate the mass of nitrifying organisms in the biofilm, and the second to calculate the mass of nitrifying organisms in the mixed liquor.

The mass balance on the biofilm can be summarized as follows:

$$\frac{d}{dt}\,Nitrifiers_{BF} = 0 = Growth_{BF} - Decay_{BF} - Sloughing$$

Whereas the mass balance on the mixed liquor is as follows:

$$\frac{d}{dt}\,Nitrifiers_{MLSS} = 0 = Growth_{MLSS} - Decay_{MLSS} + Sloughing - Wastage$$

Terms to note in these mass balances are rates of sloughing, *Sloughing*, and growth in the biofilm, $Growth_{BF}$, and in the mixed liquor, $Growth_{MLSS}$. *Sloughing* is defined as a function of nitrifiers in the biofilm, M_{BF}, and using the concept of a biofilm solids retention time SRT_{BF} as follows:

$$Sloughing = \frac{M_{BF}}{SRT_{BF}} \tag{2.4}$$

Growth in the biofilm is calculated as proportional to the ammonia flux into the biofilm. Flux, in turn, is related to bulk ammonia concentration N, a half saturation constant $k_{n,BF}$ to describe diffusive limitations in the biofilm, and the maximum achievable flux $J_{N,max}$:

$$Growth_{BF} \propto J_N = \frac{N}{k_{n,BF} + N} J_{N,max} \tag{2.5}$$

Growth in the mixed liquor is calculated as a function of the maximum specific growth of nitrifying organisms $\hat{\mu}$, the bulk ammonia concentration N, and a half saturation constant K_n to describe diffusive limitations in the mixed liquor:

$$Growth_{MLSS} = \hat{\mu}\frac{N}{K_n + N} \tag{2.6}$$

The ammonia concentration in the bulk liquid, N, is the common variable that links the mass balances for the biofilm and the mixed liquor. A solver function, such is available in an Excel spreadsheet, can therefore be used to find an iterative solution to N. And under the assumption of a complete-mix bioreactor, N is equivalent to the effluent ammonia concentration from the process.

Select parameters used in the design method of Sen and Randall, as implemented in Metcalf & Eddy, are presented in Table 2.2 [1]. These values provide valuable insight into assumptions about biofilm behavior in IFAS processes.

The design methods proposed by Sen and Randall provide a useful tool for IFAS process design. Because the methods require iterative solution for the bulk ammonia concentration, it is most appropriately implemented in a spreadsheet design tool.

2.4.5 MASS BALANCING WITH ASSUMED BIOFILM NITRIFICATION

There still remains the possibility of developing design equations based on analytical solution to the mass balance of ammonia and nitrifying organisms in a hybrid process. This would be equivalent to the design equations for conventional activated

Table 2.2
Parameters Used in Biofilm/Mixed Liquor Mass Balance Design Approach from Sen and Randall

Parameter	Value	Units	Comments
$J_{N,max}$	1.4	g N/m^2/d	Maximum potential nitrification rate at DO=6 mg O_2/L where ammonia is non-rate limiting
$J_{N,max}$	1	g N/m^2/d	Maximum potential nitrification rate at DO=4 mg O_2/L where ammonia is non-rate limiting
$J_{N,max}$	0.9	g N/m^2/d	Maximum potential nitrification rate at DO=3 mg O_2/L where ammonia is non-rate limiting
SRT_{BF}	6	days	Retention time of nitrifying organisms grown in the biofilm prior to detachment, or sloughing, in the bulk mixed liquor
$k_{n,BF}$	2.2	mg N/L	Bulk ammonia concentration at which maximum potential nitrification rate, $J_{N,max}$, is reduced by a factor of 2.

sludge presented in Equations 1.1 and 1.2 of Chapter 1. But the basis for these equations would need to account for the ammonia removal in the biofilm, as well as sloughing of nitrifiers from the biofilm to the mixed liquor. A simple equation that accounts for these factors would provide a valuable tool for engineers in developing design curves and identifying parameter sensitivities.

As will be demonstrated in Chapter 4, analytical solutions to the mass balance of nitrifying organisms in the mixed liquor can be derived if the fraction of ammonia removed in the biofilm $F_{Nit,B}$ is assumed, rather than calculated, as in the method of Sen and Randall. Design equations and curves are presented in Chapter 4 based on this approach.

2.4.6 PROCESS MODELING

Most of the commercially available process simulation software packages allow for modeling of IFAS processes. The models use a similar conceptual framework to that of Sen and Randall. There are notable exceptions, however, related to simulating biofilm nitrification rates and nitrifier sloughing rates.

Nitrification rates in the biofilm are calculated in process models from nitrifier concentrations in the biofilm, dissolved oxygen concentrations and bulk ammonia concentrations. Nitrifying organism concentrations in the biofilm are subject to out-competition from heterotrophic organisms under conditions where soluble BOD loading is high. In comparison, the method of Sen and Randall simply calculates nitrification rates based on bulk ammonia concentrations and an assumed maximum nitrification rate.

Sloughing rates of nitrifying organisms are calculated based on detachment rates from the surface of a multi-layer biofilm where nitrifying organisms typically grow in the inner biofilm layers. The biofilm SRT can vary on a case-by-case basis and the retention time of nitrifying and heterotrophic organisms in the biofilm is not the same. In comparison, the method of Sen and Randall simply assumes a biofilm SRT.

The use of simulation software for hybrid process design will be presented in Chapters 5 and 6.

2.5 OPEN QUESTIONS

Use of media-supported biofilms in activated sludge processes has not ceased to evolve since early experiments with asbestos sheets in the 1930s. This evolution reflects both the challenges and opportunities that exist for hybrid processes. Today, MBBR/AS IFAS is probably the most popular hybrid process because it effectively addresses the main shortcoming of previous technologies: biofilm thickness control.

The emergence of MABR should be seen as an attempt to solve one of the primary shortcomings of MBBR: oxygen-limited conditions in the biofilm which limit achievable nitrification rates. And operational experience to date indicates that, in this respect, the technology is successful. On the other hand, MABR may be prone to all of the challenges traditionally associated with fixed-media technologies: poor biofilm thickness control and nuisance growth of red worms. Moreover, the cost of MABR media, per m^2 of surface area, is higher than MBBR media. This could change with time as MABR technologies are widely adopted and production of media achieves economies of scale. Nevertheless, someone might reasonably ask: Do the higher nitrification rates achievable with MABR justify the higher cost per m^2 of surface area?

It may be too early to judge the shortcomings of MABR technology identified above, and whether they can be suitably mitigated through design features and good process design. It is expected that the types of media and equipment, as well as the knowledge and tools available for process design, will continue to evolve in the coming years to address new challenges and opportunities as they arise. In anticipation of this, some of the open questions that remain for hybrid systems are identified below.

Impacts of BOD Loading on Biofilm Nitrification

The impacts of BOD loading on nitrification in MBBR biofilms has been studied fairly extensively and, typically, placement of media in the upstream portions of the bioreactor is not recommended. How are nitrification rates in MABR biofilms impacted by BOD loading? And how does this limit placement of MABR media in a bioreactor?

Implications for Ammonia- and Oxygen-Limited Biofilms

MBBR flux curves indicate that ammonia is not limiting to nitrification until bulk liquid concentrations drop below 0 to 1 or 2 mg N/L [11]. This is only a marginally higher range than for mixed liquor nitrification which might be expected to be ammonia-limited in a range of 0 to 0.5 mg N/L.

Data from piloting indicates that nitrification in MABR biofilms becomes limiting at a bulk liquid ammonia concentration in a range of 5 to 10 mg/L [19]. Data from a full-scale plant will also be presented in Chapter 6 that indicates biofilm nitrification being ammonia-limited between 0 and 20 mg N/L. Compared to MBBR media, this is a very broad range. Although oxygen abundance in the MABR biofilm is an important reason for this behavior, an additional factor that must be considered is mass transfer limitation between the inside of the cassettes and the surrounding bulk liquid.

So what are the process design implications of a broad (MABR) or narrow (MBBR) range of ammonia-limited conditions? As discussed in Chapter 1, ammonia-limited conditions provide the ability to treat peak ammonia loads, which is one of the main reasons for applying safety factor to mixed liquor SRT. The consequences of this are explored in greater detail in Chapter 6, "Investigation of Operational Dynamics".

Maximum Packing Densities

Practical experience has shown that a maximum fill fraction of about 60% can be achieved in an MBBR/AS process and that prudent design targets something closer to 40 to 50%. This is relatively easy to test and correct in an MBBR/AS process due to the mobile nature of the media. If the bioreactor tank contains too much media then the excess can simply be removed. But what is the maximum packing density that can be achieved with Fixed-media, and MABR media? And how is this impacted by BOD loading and the energy applied for biofilm scouring? At a certain point the void ratio becomes prohibitively small and growth of even a thin biofilm is likely to turn the media frames into a solid mass of sludge. For Fixed-media and MABR media, packing density is selected during product design and, unlike MBBR media fill fraction, cannot easily be corrected in the field.

Shortcut Nitrogen Removal

What is the potential for shortcut nitrogen removal in hybrid processes? Can specially-tuned aeration strategies in the MABR media create conditions to suppress NOB growth and even promote anammox growth? Researchers have highlighted the unique opportunities of MABR biofilms for shortcut nitrogen removal, but these still need to be proven at full-scale [23].

Role of Higher Life Forms

Higher life forms including protozoa, rotifers, amoeba and nematodes (red worms) can be abundant in biofilms. Despite this, commonly used process models and design equations only explicitly account for the activity of bacteria.

Using modeling approaches developed for activated sludge mixed liquor, it is assumed that the activity of higher life forms is captured in standard growth and decay parameters. However, is this really a valid assumption for biofilms where higher life forms are more abundant? Does their grazing and predation activity impact assumptions about achievable flux rates or nitrifier sloughing from the biofilm to the mixed liquor? Is grazing activity very different in co-diffusional *vs.* counter-diffusional biofilms? It has been shown, for example, that amoeba and protozoa proliferate in the interior, aerobic layers of counter-diffusional (MABR) biofilms, and their grazing activity can significantly increase void spaces [2]. One effect of this could be to limit the growth potential of nitrifying organisms, which also tend to grow at the base of biofilms. Another effect could be to increase the rate of biofilm sloughing. The former effect would be detrimental in a hybrid process, but the latter could be positive. More research into the role of higher life forms in biofilms should provide valuable insight into how to improve hybrid system designs, and the relative potential of co-diffusional and counter-diffusional biofilms.

3 Hybrid Systems in Practice

Design approaches for hybrid systems presented in the previous chapter evoked concepts such as the "equivalent SRT" and "seeding effect" to describe the synergistic benefits of operating a hybrid activated sludge process. The purpose of this chapter is to investigate how this plays out in practice. Does adding media-supported biofilm to the activated sludge process provide any benefit in the real world? Does it really translate into process intensification? And do the savings in capital and operating costs offset the cost of media and installation?

This chapter focusses firstly on hybrid IFAS processes which are designed to nitrify below conventional design guidelines. In short, we will review the process intensification claims of IFAS. The discussion is then extended to include discussion of the seeding effect in TF/AS and sidestream/mainstream processes. These processes, though not hybrid processes in the strict sense as defined in Chapter 2, still experience the seeding effect. Indeed, a better understanding of the seeding effect in IFAS can be had by looking at processes where the source of the seeding effect is external to the activated sludge process itself.

The theoretical framework for understanding and quantifying the seeding effect will be further explored in Chapter 4 to develop a series of design equations. For this chapter, the goal is to provide an overview of the following:

- Drivers for implementation of MBBR/AS IFAS and operational experience from two full-scale installations.
- Drivers for implementationof MABR/AS IFAS and operational experience from one full-scale and one pilot demonstration installation.
- Experience from activated sludge processes when bioaugmentation, of nitrifier seeding, is from an upstream or parallel process.

3.1 BALANCING BIOFILM AND MIXED LIQUOR NITRIFICATION

3.1.1 NITRIFYING ENTIRELY IN THE BIOFILM

In hybrid processes, where the required amount of nitrification can be achieved entirely in the biofilm, the synergistic relationship between biofilm and mixed liquor nitrification need not be of much concern. In such cases, the mixed liquor can be be comfortably operated at the minimum SRT required to satisfy the faster growing heterotrophic organisms, the objective being to achieve good flocculation of these organisms which will then lead to good BOD and TSS removal. This usually means a minimum SRT of 2 to 3 days. This would be the ultimate case of treatment intensification: complete decoupling of the mixed liquor SRT from requirements for nitrification.

To illustrate, consider the "baseline alternative" for upgrading a non-nitrifying plant to nitrification. The existing plant operates at 3 days SRT, which is insufficient to provide reliable nitrification at winter temperatures of 10°C. To meet a newly introduced effluent ammonia limit, and to be within acceptable design guidelines, the process would likely need to operate anywhere from 9 to 15-day SRT. The baseline upgrade scenario would consider that the aeration tank and clarifier volumes need to be at least tripled or risk a significant capacity derating. If one considers that the cost of building new tank volume can be on the order of 4 USD per gallon, the capital cost implications of the new permit are serious to say the least!

In contrast, an IFAS solution where all of the required nitrification occurs in the biofilm would require construction of no additional tankage. The capital cost of upgrading from a non-nitrifying to a nitrifying plant, while maintaing the existing treatment capacity, would be the cost of biofilm support media and any ancillary equipment. While not negligible, this cost would certainly be less than the cost of building new bioreactor tanks and clarifiers: it is not hard to see how the IFAS upgrade would be the preferred scenario.

As will be seen in the case studies discussed later in this chapter, IFAS designs tend to rely on a portion of the nitrification taking place in the biofilm, and the remainder taking place in the mixed liquor. There is a range of factors that may prevent it from being practical to achieve all of the required nitrification in the biofilm. So the synergistic effects of biofilm and mixed liquor nitrification, and indeed the seeding effect, cannot be ignored.

3.1.2 PRACTICAL CONSTRAINTS ON BIOFILM NITRIFICATION

The following points are discussed within the context of retrofitting existing activated sludge bioreactors for IFAS. They explain why it is often impractical to achieve 100% of the nitrification in the biofilm. This is not to say that it is an impossible goal. After all, this is how "pure biofilm" MBBR processes are designed. However, pure biofilm processes are typically designed in two-stage configurations, one stage for BOD removal and the second for nitrification, or as a tertiary stage to an existing activated sludge process. Also, purpose-built tanks would typically be used that could be designed with correct volume and dimensions to achieve the desired packing density and avoid hydraulic bottlenecks. For activated sludge process retrofits, there are often practical limitations to achieving the desired quantity of biofilm support media, and designers have to adjust their designs to work within the constraints of the existing bioreactors.

The Effects of BOD Loading

Placement of media at the upstream end of bioreactors is often avoided where, due to high BOD loading, it may not be possible to achieve a nitrifying biofilm. Under high BOD loading, heterotrophs in the biofilm can outcompete nitrifying organisms for the available dissolved oxygen. As a result, IFAS design tends to place the media towards the downstream end of the bioreactor where most, or all, of the influent BOD has already been removed by the mixed liquor organisms. Nitrification in MABR biofilms seems to be less sensitive to BOD loading than MBBR biofilms due to the special properties of the "counter-diffusional" biofilm they support. This was discussed in Chapter 2. As a result, current design practice for MABR/AS involves placing media near the front end of the bioreactor. This will be discussed in the MABR/AS case study presented in Section 3.3.

Biofilm Packing Density

There are practical limits to the volumetric surface area that can be achieved, *i.e.* the m^2 of media surface area per m^3 of bioreactor volume. Mobile media provide the highest surface areas but are limited by the volumetric fill ratio of media that can be achieved in the bioreactor tank. This ratio represents the m^3 of media per m^3 of bioreactor and in most cases should not exceed 50%. Above this ratio, mixing of the media becomes compromised and in the words of one operator "the media starts crawling up the walls".

As an example, assuming a specific surface area of 500 m^2 per m^3 of mobile MBBR media, and a 50% fill fraction, a tank surface area of 250 m^2 per m^3 of bioreactor volume is calculated. For fixed media, including MABR media, packing density is the metric analogous to the MBBR "media specific surface area". Taken together with the tank fill ratio, the fixed media packing density can be used to calculate the tank surface area. Typically, the result is a tank surface area approximately half of what can be achieved with MBBR media.

Hydraulic Bottlenecks

In the case of MBBR media, the need to install retention sieves in the bioreactor adds a significant head loss to the hydraulic gradeline. This can limit the flow capacity through the process and can, in itself, make the MBBR option impractical. Such considerations do not apply to Fixed-media and MABR which, because they are fixed, do not require media retention sieves.

Diffusion Resistance

To reliably meet an effluent ammonia limit, even one as high as 5 mg N/L, a designer should target a much lower operating objective, as low as 1 or 2 mg N/L. Nitrification in biofilms is a diffusion limited process and bulk liquid ammonia concentrations below 5 mg N/L will significantly impact achievable rates inside the biofilm. To ensure that the biofilm can remove all of the ammonia load to an objective of less than 5 mg N/L, a designer would need to assume a low nitrification rate. This can be compensated for by increasing media surface area; however, this may not be achievable based on maximum fill ratios or packing densities in the bioreactor.

3.1.3 BIOFILM AND MIXED LIQUOR NITRIFICATION

Due to these factors, and others that are often more site specific, IFAS designs tend to rely on some portion of the nitrification still taking place in the mixed liquor. But there would be no benefit if they still maintained a traditional nitrification design SRT. As will be shown in the following sections, representative IFAS processes in the U.S. operate at SRTs in the range of 2.5 to 5 days with temperatures from 8 to 18°C. These plants are identified by Ødegaard (2009) as operating at "60% of what would be needed by activated sludge alone" relative to German ATV design standards [36]. Ødegaard further identifies that at least one, if not two, of these plants operate below the washout SRT as determined by nitrifier growth kinetics.

So what is the synergistic effect of hybrid biofilm/AS systems that allow them to achieve mixed liquor nitrification at lower than required SRTs? Bioaugmentation, or more typically the "seeding effect", is the response provided in papers and design manuals [36, 1, 9]. However, another possibility is that, when most of the ammonia load is already removed on the biofilm, it makes it easier for the mixed liquor to remove what is left. This would seem to contradict the traditional design equation presented in Equation 1.1 which indicates that effluent ammonia concentration is only dependent on SRT and nitrifier kinetics, not load to be treated. As will be shown in Chapter 4, however, this assumption is not entirely valid, particularly when considering the seeding effect.

Another possibility that must be considered is that these plants are simply operating at a reduced safety factor. Except for the one or two plants that can be demonstrated to actually be operating below the washout SRT, it may be that these other plants tolerate a lower safety factor due to:

- a more relaxed effluent ammonia permit than that considered when developing the ATV design guidelines cited by Ødegaard (2009) [36],
- site specific conditions which would reduce the likelihood of effluent breakthrough events, such as less influent load variability or greater process redundancy,
- a greater tolerance for risk on the part of the operator or process designer.

The following sections review case studies of conventional activated sludge processes that were retrofit as IFAS processes: both MBBR/AS and MABR/AS. The drivers for the process retrofit are presented as well as operational experience and evidence that the seeding effect is enabling operation at reduced SRTs.

3.2 MBBR/AS CASE STUDIES

3.2.1 BROOMFIELD, CO WWTP

Project Drivers

The first MBBR/AS plant in the U.S. was commissioned in the City of Broomfield, CO in 2002 and 2003 as a 20,000 m^3/d facility with a population equivalent

(PE) of approximately 50,000. The drivers for the upgrade were both population increase and a new effluent ammonia treatment objective of 1.5 mg N/L in summer, and 3 mg N/L in winter. In addition, to enable reuse of a portion of the wastewater for irrigation, there was an objective to remove total inorganic nitrogen to less than 10 mg/L and total phosphorus to less than 1 mg/L. Effluent objectives for both BOD_5 and TSS of 10 mg/L needed to be met at the discharge of the secondary clarifiers as the plant does not benefit from a tertiary treatment stage. The above design criteria were developed based on 30-day averages. Six alternative treatment strategies were evaluated for the required upgrade based on the criteria of compatibility with future expansion, similarity with the TF/AS process, footprint requirements and overall cost. The MBBR/AS IFAS technology was selected based on these criteria [27].

Process Configuration

To meet the effluent TIN and TP requirements, an A^2O process configuration was designed with a small pre-anoxic zone to receive primary effluent and the RAS, followed by a larger anaerobic zone to promote enhanced biological phosphorus removal. The anaerobic zone is followed by an anoxic zone for denitrification which receives a nitrified mixed liquor recycle from the downstream aerobic zone. The aerobic zone is made up of two cells, each designed with a 30% fill fraction of Kaldnes K1 media. The K1 media has a specific surface area of 500 m^2/m^3. The relative volume fractions of the pre-anoxic, anaerobic, anoxic and aerobic zones is 5%, 9%, 19% and 67%, respectively.

Operating SRT

The design SRT was 4.7 days at 13°C which, given the significant unaerated volume fractions, gives an aerobic SRT of 3.15 days. During a two-year evaluation period, Rutt *et al.* (2006) describe the plant operations in a range of 3 to 4 days aerobic SRT at winter temperatures of 14°C while consistently maintaining an effluent ammonia concentration less than 1 mg N/L [27].

Using Equation 1.2, and at a temperature of 14°C, the minimum SRT to achieve an effluent ammonia of 1 mg N/L is estimated at 3 to 4 days. As such, in this conventional activated sludge design paradigm, the IFAS process at Broomfield is operating right at the minimum SRT, *i.e.* with a safety factor of 1. The reliability of operations at Broomfield indicate that the IFAS safety factor must in fact be higher than this. Chapter 4, Section 4.3.4, provides a novel means for quantifying this IFAS safety factor using the ratio $R_{Nit,max/L}$. This ratio accounts for the nitrifying activity in the mixed liquor, and the fraction of ammonia load removed in the biofilm.

The fraction of the ammonia load removed on the biofilm was not measured during this study but some insights can be gained from looking at the ammonia loading rate relative to the media surface area. Given the influent max month ammonia load of 1,126 kg N/d, aerobic zone volume of 4,546 m^3/d, a media fill fraction of 30%, and a Kaldnes K1 media specific surface area of 500 m^2/m^3 we can calculate a 1.65 g N/m^2/d. Current practice is to design for an MBBR media nitrification rate of less than 1 g $N/m^2/d$ and so it can be concluded that the fraction of the ammonia load removed in the biofilm was considerably less than 100%.

Biomass Coverage

Biomass coverage on the media was consistently higher in the first MBBR cell (3-20 g/m62) than in the second (1-10 g/m^2) but also varied considerably on a seasonal basis. Higher biomass coverages were noted to occur in winter and lower coverages in summer. No correlation with mixed liquor was found, which operated in a range of 2,700 mg/L to 1,500 mg/L, during the two-year study period. Settling properties in the mixed liquor were good during this period with SVI's in the range of 100-150 mL/g. The consistency of process performance despite highly variable media coverage confirmed the design assumption that effective surface area is the critical criteria to performance, not media coverage [27].

3.2.2 JAMES RIVER TREATMENT PLANT

Project Drivers

The James River Treatment Plant is one of nine major treatment plants operated by the Hampton Roads Sanitary District (HRSD) in southeast Virginia. The plant has a rated capacity of 17 MGD (64,000 m^3/d) and discharges into the James River, a major tributary to the lower Chesapeake Bay. The Chesapeake Bay is sensitive to nitrogen loading and dischargers to its watershed, including HRSD, have been subject to increasingly stringent nutrient limits.

Meeting more stringent effluent TN requirements at the James River Treatment Plant meant introducing unaerated zones at the front end of the bioreactor for denitrification. This reduces the aerobic portion of the SRT, the nitrification capacity, and would have derated the capacity of the plant without the IFAS retrofit that was undertaken to restore the nitrification capacity.

Process Configuration

Figure 3.1 presents the process flow diagram for a 2 MGD demonstration scale test run from November 2007 to April 2009. The purpose of the testing was to explore IFAS performance and investigate the technology as an option for the full-scale plant upgrade [33]. The testing was successful and provided the basis for converting the entire plant to MBBR/AS IFAS.

The test conducted at James River Treatment Plant was noteworthy due to the amount of complementary testing that was conducted to support the performance evaluation. This included bench-scale activity testing of media sampled from the plant, and model calibration of the full-scale and bench-scale results. In addition, it is important to note that good control of the mixed liquor SRT was possible due to the use of hydraulic wasting of mixed liquor from the bioreactor effluent. With hydraulic wasting, mixed liquor SRT is only a function of the waste activated sludge (WAS) volume and does not depend on WAS concentration. In constrast, calculating SRT when wasting is from the return activated sludge (RAS) stream requires measurement

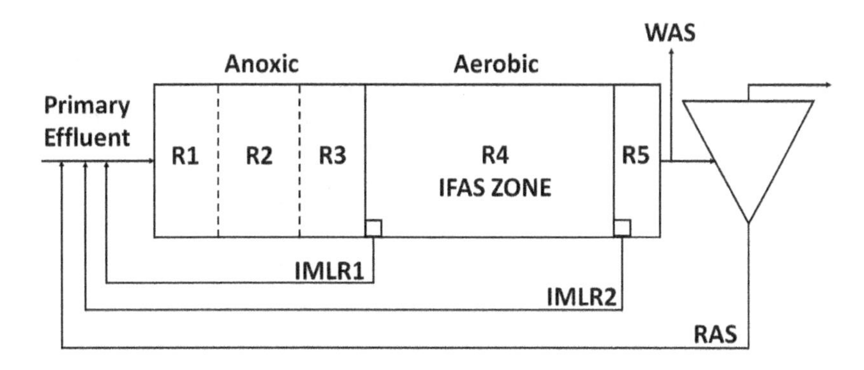

Figure 3.1 Process flow diagram of MBBR/AS IFAS process at James River

or estimation of RAS/WAS concentration, which can introduce uncertainty into the SRT estimation.

Key findings for the James River Treatment Plant IFAS testing are summarized below based on the work of Thomas (2009) [33].

Nitrifying Below the Washout SRT

Effluent ammonia concentrations of 0.5 mg N/L were achieved in winter at a mixed liquor SRT of 2.9 d and a minimum temperature of 14.5°C. The minimum conventional SRT for this temperature and condition was estimated at 3.5 d and the percent removal of ammonia in the biofilm *vs.* the mixed liquor were estimated at 75% and 25%, respectively. The IFAS demonstration testing at the James River Treatment Plant validated nitrification below the washout SRT with 25% of the nitrification taking place in the mixed liquor.

Carrier Biomass Content

The carrier biomass content refers to the biofilm attached to the media as measured by scraping biofilm off of the media and drying in an oven at 105°C. It is easier to measure than biofilm thickness, which cannot practically be measured *in situ* in its natural, buoyant state. Arguably, it is also a more fundamental parameter being independent of biofilm density, which may vary based on temperature and biomass composition. Carrier biomass content was found to be variable with respect to temperature and mixed liquor SRT. Observations were in line with typical experience in IFAS processes in that carrier biomass content was highest in winter.

Nitrification in Biofilm vs. Mixed Liquor

It was observed that nitrifying activity shifted from the biofilm to the mixed liquor as water temperatures increased, and back from the mixed liquor to the biofilm as temperatures decreased. During the hottest periods, as little as 30% of nitrifying activity was observed in the biofilm. In contrast, 75% of nitrifying activity was observed in the biofilm during the coldest winter periods. This indicates that as washout SRT is approached under winter temperatures, the hybrid process compensates by shifting nitrification activity to the biofilm. The biofilm provides added safety factor or resilience to loss of nitrification in the mixed liquor.

Ammonia Oxidizing vs. Nitrite Oxidizing Activity

AOB and NOB activity did not segregate itself evenly between the biofilm and mixed liquor. NOB activity in the biofilm remained consistently higher than AOB activity in both cold and warm weather conditions. This finding is significant for potential considerations around shortcut nitrogen removal, in which achieving NOB suppression plays a vital role. Higher NOB activity in the biofilm suggests that achieving shortcut nitrogen in IFAS processes may be more challenging, if not impossible. Considerations around shortcut nitrogen removal aside, these results could also be viewed positively in that increased NOB activity in biofilms could help avoid the risk of nitrite in the effluent. Discharging nitrite in the effluent is undesirable both for its toxicity effects and its potential impacts on downstream disinfection processes.

3.3 MABR/AS CASE STUDIES

3.3.1 YORKVILLE-BRISTOL SANITARY DISTRICT

The Yorkville-Bristol Sanitary District (YBSD) owns and operates a wastewater treatment plant in the City of Yorkville, Illinois, serving a population of approximately 18,500. The plant has a rated capacity of 13,700 m^3/d and an organic load of 2,155 kg/d of five-day biochemical oxygen demand, BOD_5. The plant effluent discharges into the Fox River. A plan view of the plant is presented in Figure 3.2.

As part of broader upgrade to a biological nutrient removal process, MABR media was installed in the bioreactor to offset the loss of nitrification capacity associated with moving from a 100% to a 60% aerobic volume fraction. The resulting MABR/AS IFAS process is the first full-scale facility of its kind in the U.S. and much is still being learned about process operations and performance. The sections below summarize information and learnings from a recent paper given by Underwood *et al.* on the startup of the MABR/AS process at YBSD [34].

Drivers for Upgrade

The drivers for the MABR/AS upgrade at YBSD relate both to the need for increased capacity and new, more stringent effluent treatment objectives, so more quantity and better quality. According to Underwood *et al.* (2018), YBSD has experienced an increase in organic load due to population growth and new industrial contributions. However, and in line with broader trends around the world towards water conservation, hydraulic loading has not increased at the same rate. Nevertheless, because the State of Illinois uses organic loading limits based on the Ten States Standards, the anticipated increased organic load will result in the plant approaching its rated organic capacity.

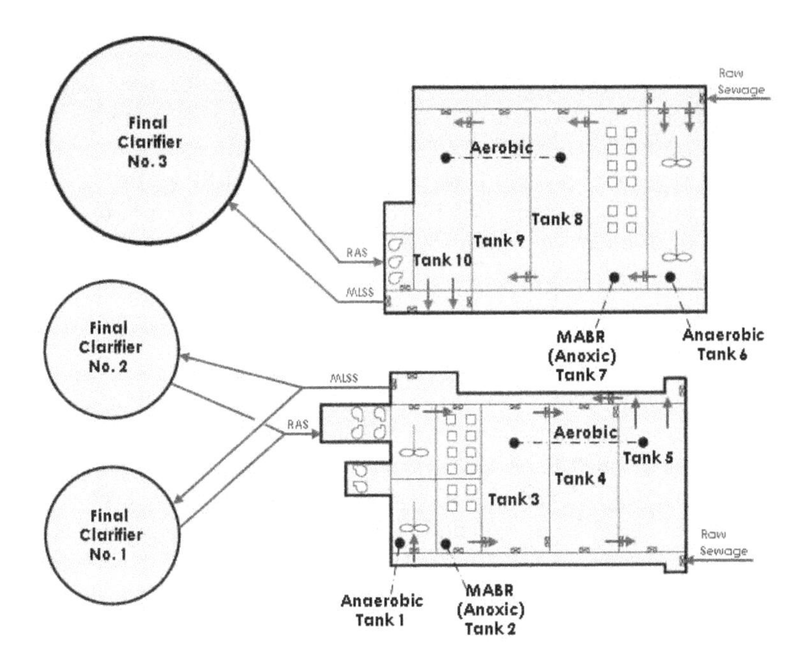

Figure 3.2 Process flow diagram of MABR/AS process at YBSD, courtesy of YBSD

At the same time, YBSD must comply with a new total phosphorus (TP) effluent limit of TP less than 1.0 mg/L by May 2019 under the plant's National Pollutant Discharge Elimination System (NPDES) permit. This situation requires the plant to make modifications to increase organic treatment capacity and implement phosphorus removal. The existing plant footprint is built-out and any increase in conventional treatment capacity would require construction of a separate treatment plant on adjacent property. YBSD was interested in an alternative solution that would minimize construction and capital expenditures to limit the impact on the YBSD rate payers while accelerating the implementation schedule by avoiding the time required for permitting and construction of a new treatment plant.

YBSD chose to retrofit the existing secondary treatment process to an enhanced biological phosphorus removal (EBPR) and Membrane Aerated Biofilm Reactor (MABR) system in the existing activated sludge reactors for the following reasons:

1. This would enable balancing of the organic capacity and the hydraulic capacity of the plant. YBSD could increase the organic capacity of the plant while maintaining the current hydraulic capacity and thereby obtain more treatment capacity out of its existing assets.
2. The capital cost of the MABR and EBPR solution was estimated to be less than 25% the cost of building a new conventional treatment plant.

3. The MABR and EBPR solution could be implemented within eighteen months.
4. Synergy between the MABR and EBPR processes exists as the simultaneous nitrification and denitrification (SND) in the MABR decreases the nitrate load in the return activated sludge (RAS) to the anaerobic zone. Less nitrate in the RAS is a key strategy for improving EBPR performance.
5. The energy efficiency of the MABR process was expected to result in no net increase in energy consumption for the plant, even with the increased organic loads.

Lessons Learned

As reported by Underwood *et al.*, the performance of the ZeeLung media has been very stable with average oxygen transfer rates (OTR) and oxygen transfer efficiencies (OTE) of 10.8 g $O_2/m^2/d$ and 33.3%, respectively. Assuming an ideal stoichiometric ratio of oxygen utilization to nitrification rate of 4.57 g O_2 per g N, this would equate to a nitrification rate (NR) of 2.36 $g/m^2/d$, which is in line with data reported from other MABR references.

The presence of a nitrifying biofilm was corroborated using quantitative polymerase chain reaction (qPCR) analysis of DNA/rRNA in the ZeeLung biofilm that showed AOB and NOB represent approximately 40% of the total bacteria. In contrast, qPCR analysis of activated sludge mixed liquor typically reveal AOB and NOB significantly less than 10% of the total bacteria.

So qPCR results reveal a higher than normal abundance of nitrifying organisms in MABR biofilms. Nevertheless, the biofilm is still composed of more heterotrophic than nitrifying organisms. This does not contradict claims that the majority of the oxygen in the biofilm is used for nitrification, however. The reason for this is that the oxygen utilization associated with the growth of nitrifying organisms is much higher than it is for heterotrophic organisms. As illustrated in Figure 3.3, the ratio of oxygen utilization to growth for nitrifiers is $4.57/Y_{Nit}$ g O_2/g biomass whereas it is $(1\text{-}Y_H)/Y_H$ for heterotrophic organisms. If one assumes a yield of nitrifying organisms ($Y_{AOB}+Y_{NOB}$) of 0.24 g COD/g N and a yield of heterotrophic organisms of 0.67 g COD/g COD, then the ratios of oxygen utilizations to growth, in g O_2 per g COD growth, are 19:1 and 0.5:1 for nitrifiers and heterotrophic organisms, respectively.

The higher oxygen utilization to growth ratio, 19:1 *vs.* 0.5:1, makes it clear how nitrifying organisms, despite comprising less than half the biofilm composition, can be responsible for almost the entirety of the oxygen utilization. Moreover, a portion of the heterotrophic activity in the biofilm is likely denitrifying activity: using nitrate generated by nitrifying organisms as the electron acceptor instead of oxygen.

Initial results of the MABR/AS at YBSD were promising, showing that the process is compatible with EBPR when operated as an A^2O process with a 20% anaerobic volume fraction. Results from the MABR/AS process at YBSD will be further explored in Chapter 6.

Figure 3.3 Comparison of stoichiometry of nitrification compared to heterotrophic organism growth

3.3.2 UK MABR/AS DEMONSTRATION PILOT

Project Drivers

Sunner *et al.* (2018) describe operations of a full-year demonstration pilot of the hybrid MABR/AS process at a WWTP in the southern United Kingdom between March 2017 and April 2018 [31]. The UK has been subject to significant population growth concentrated in this region. This has resulted in the need to extend treatment works capacity in areas where land is at a premium and where site footprints are tight. Population increase has also led to more stringent treatment consents being enforced. Population growth has focussed attention on both the "quantity" and "quality" aspect of sewage treatment: plants in this part of the UK need to increase their rated capacity as well as improve their treatment performance. As has been discussed, increasing treatment quality normally comes at the expense of rate capacity.

The only solution to the quality/quantity conundrum within constrained site footprints is to intensify the existing technologies. This is the context in which the investigation of the MABR/AS process was undertaken. MABR/AS was selected as the IFAS approach based on technical and economic evaluation that showed it could provide the lowest whole life cost for the upgrade of two candidate activated sludge plants [31].

Pilot Overview

The process flow diagram for the pilot is presented in Figure 3.4 where ZeeLung MABR media is installed in approximately 15% of the bioreactor volume. The two configurations investigated were:

- Configuration A: Installation of MABR media in an anoxic zone immediately following a small pre-anoxic zone of 7% volume fraction.
- Configuration B: Installation of MABR media in an anoxic zone downstream of an aerobic zone with 18% volume fraction.

Configuration A is reminiscent of the YBSD process configuration wherein the MABR is located after an initial unaerated zone. The purpose of this initial zone is to remove influent readily biodegradable BOD that could stimulate heterotrophic

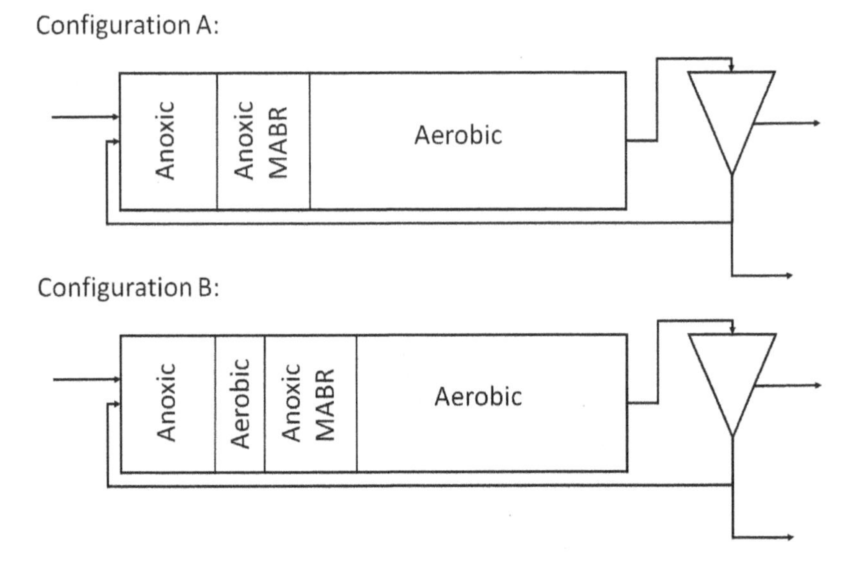

Figure 3.4 Process flow diagram from Sunner *et al.* for the UK MABR/AS pilot as operated in Configurations A and B

growth in the biofilm, at the expense of nitrification. Configuration B was tested to mimic designs of MBBR/AS where media is further located downstream in the process after an initial aerobic contact zone. This should have allowed for greater removal of BOD than was achieved in the pre-anoxic zone from Configuration A. As it happened, no significant differences in nitrification rate were observed between the two configurations indicating that BOD loading was not impacting nitrification rates in Configuration A.

The pilot operated at a feed flow rate of approximately 5 m^3/d of primary effluent, a mixed liquor SRT typically in the range of 2 to 10 d, and included an MABR cassette with ZeeLung MABR media surface area of 40 m^2. The pilot achieved an average of 96% removal of influent ammonia of which 21 to 34% was removed in the biofilm.

Pilot Performance

There is already considerable experience in the UK with IFAS technologies and process design procedures are well established. For this reason, pilot testing focussed mainly on demonstrating the unique claims of the ZeeLung MABR technology, namely higher nitrification rates and lower energy requirements as compared to conventional Fixed-Media/AS and MBBR/AS IFAS. As such, the performance targets were:

1. A nitrification rate (NR) of 2.0 g/m^2/d where surface area was a single 39 m^2 ZeeLung module.

2. Removal of 20% of the influent ammonia load in the biofilm, *i.e.* $F_{Nit,B}$ = 20%.
3. An aeration efficiency (AE) of 4 kg O_2/kWh: the AE of the ZeeLung media was calculated based on the ratio of the oxygen transfer rate (OTR) to the total energy inputs for the operation of the MABR zone, including the process air to the media and the air for mixing and scouring of the module.

Each of these performance metrics was met with average values of NR, $F_{Nit,B}$ and AE over the course of the piloting, respectively, of 2.2 to 2.7 g N/m^2/d, 21 to 33% and 4 to 4.9 kg O_2/kWh. In addition, performance of the overall hybrid process was verified with respect to typical effluent treatment consents for the UK:

- Effluent ammonia less than 2 mg N/L
- Effluent BOD_5 less than 10 mg/L
- Mixed liquor SVI less than 150 mL/g (Note: SVI was used as a surrogate for effluent TSS because of challenges in operating a pilot-scale secondary clarifier.)

Effluent ammonia as measured using composite samplers was consistently below 0.5 mg N/L with short periods of ammonia breakthrough, to as high as 5 to 7 mg N/L, occurring as a result of low dissolved oxygen during peak loading periods, *i.e.* "DO sag". DO sag during periods of high loading was an issue in this pilot due to an insufficient air supply to the aerobic zones [31].

The exceedences due to DO sag events are reflective of the challenges of operating at pilot scale where process configurations and conditions are often subject to multiple changes and adjustments. As a result, equipment is not always "right sized" for every experimental phase. It is the expectation that operations would be more stable at full-scale where there would be sufficient blower capacity for the fine bubble aeration zone under all loading conditions. As a result, the exceedences during the DO sag events did not unduly influence the evaluation of the process as a whole.

Process Intensification

The seeding effect was demonstrated during two separate periods lasting 55 days and 42 days, respectively, when the pilot was operated at average temperatures of 19 and 12°C and average aerobic SRTs of 1.9 and 4.0 days. By comparing the washout SRT along with the operating aerobic SRT and effluent ammonia, Sunner *et al.* demonstrated process intensification when the percent ammonia removal in the biofilm, $F_{Nit,B}$, was 20 to 34%. This is considerably lower than the biofilm $F_{Nit,B}$ of 75% relied on for process intensification in the James River MBBR/AS, Section 3.2.2. In contrast, the biofilm $F_{Nit,B}$ at the Broomfield WWTP, Section 3.2.1, was not quantified but was likely significantly less than 100%.

This raises the question as to what minimum ammonia removal in the biofilm, $F_{Nit,B}$, is required to achieve meaningful process intensification? And is the intensification observed simply due to a lower ammonia load on the mixed liquor, bioaugmentation/seeding effect, or some form of dynamic load balancing provided by ammonia-limited conditions in the biofilm? The MOP35 design guidelines discussed in Chapter 2 provide a partial response to these questions but for fairly limited conditions [9]. Process modeling will be used to explore this further in Chapters 4, 5 and 6.

3.4 LEARNINGS FROM IFAS CASE STUDIES

Intensification

The review by Ødegaard of MBBR/AS processes in the U.S. indicate that the majority are achieving their nitrification objectives below conventional design SRTs [36]. The Broomfield, WWTP and James River Treatment Plant case studies make clear that both the biofilm and mixed liquor are responsible for this performance. That is to say, that nitrification in the mixed liquor is maintained in hybrid processes even at mixed liquor SRTs that would normally preclude this.

Bioaugmentation, or the seeding effect, is often cited as the explanation for mixed liquor nitrification in hybrid processes operated at low SRTs. But alternative explanations could be related to the lower ammonia load that needs to be nitrified by the mixed liquor, the load balancing effects of ammonia limitation in the biofilm, or simply operation of these plants at a reduced safety factor. A reduced safety factor may in fact be a factor in some cases, but cannot explain the performance of hybrid processes that successfully operate below conventional washout SRTs.

Biofilm/Mixed Liquor Synergies

Whatever the case, the synergistic relationship between nitrifying organisms in the biofilm and mixed liquor in IFAS systems cannot be denied. For example, seasonal shifts in the proportion of nitrification taking place in the biofilm, and in the mixed liquor, was empirically demonstrated at the James River Treatment Plant. On a diurnal basis, this same shift can be observed through the response in oxygen transfer in an MABR biofilm. How this dynamic load balancing between biofilm and mixed liquor nitrification translate into actual benefits will be explored in Chapter 6.

MABR, the New Kid on the Block

Despite its newness to the market, we see from the UK experience that MABR/AS is being considered as an IFAS upgrade option in the same manner as Fixed-Media/AS and MBBR/AS, that is to say, as just another IFAS process. The technologies are differentiated by consultants on how much tank volume is required to achieve the desired surface area and nitrification rate, and unique features such as operating energy requirements, whether media is fixed or mobile, screening requirements, and of course the cost of the media.

Outstanding Questions

It is useful to list some of the questions that have arisen from a review of the IFAS case studies in the previous sections. These questions will be addressed using the process modeling tools provided in the following chapters:

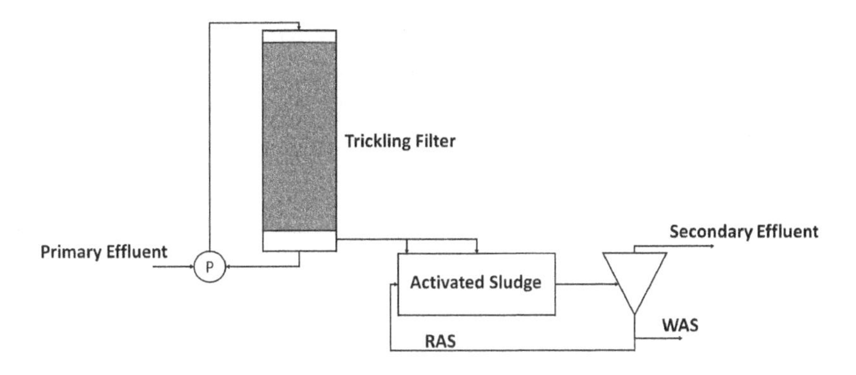

Figure 3.5 Process flow diagram for the trickling filter/activated sludge process

- What percent ammonia removal in the biofilm, $F_{Nit,B}$, is required to provide meaningful process intensification?
- What are the relative benefits of seeding effect, reduced ammonia load on the mixed liquor, and dynamic load balancing?
- Are the above effects substantially different in MBBR/AS *vs.* MABR/AS processes due to the co-diffusional *vs.* counter-diffusional natures of their biofilms?

3.5 EXTERNAL SEEDING EFFECT

3.5.1 TF/AS PROCESS

One of the first documentations of the seeding effect was from Trickling Filter/Activated Sludge (TF/AS) processes in which nitrification was observed in the activated sludge mixed liquor "at solids residence times that would otherwise preclude nitrification" [5].

The process flow diagram for a TF/AS process is presented in Figure 3.5 showing primary effluent being pumped into a trickling filter tower whose effluent flows into an activated sludge process. This configuration is representative of a number of plants including the Rowlett Creek and Duck Creek WWTPs operated in the City of Garland, Texas. In such a design, neither the trickling filter stage nor the activated sludge process is adequately designed for nitrification on their own. Trickling filters typically require a first stage for BOD removal so that nitrification can be achieved in the second stage. At the BOD loading experienced in these plants, only partial nitrification was being observed.

Full nitrification was observed in the activated sludge process despite the fact that the traditional design model (Equation 1.1) indicated that it should not. An important feature of Equation 1.1 is that it assumes no influence from either (a) influent ammonia or (b) influent nitrifying biomass. Of course, the first assumption is no longer valid below the washout SRT in which case the effluent ammonia must be a direct function of the influent ammonia, *i.e.* what comes in must go out. The second assumption is also not valid when the influent nitrifying biomass is not negligible and the operating SRT is near to, or below, the washout SRT.

This model deficiency was addressed by Daigger *et al.* by proposing an extension to the conventional design model that includes the effect of both influent ammonia and nitrifying biomass concentration [5]. Unlike influent ammonia concentrations, influent nitrifying biomass cannot be readily measured, so the authors instead proposed to infer influent nitrifying biomass concentrations from the ammonia removal in the upstream trickling filter (TF) stage, and an assumed biomass yield. The biomass yield used by the authors was described as the "true growth yield" and so presumably accounts for both the growth yield and the effect of endogenous decay in the TF biofilm. This book will refer to this yield as the "sloughed" yield, $Y_{Sloughed}$. Other references may refer to this as the observed yield, Y_{obs}, and show it as a function of the growth yield, biofilm SRT, and decay rate as follows:

$$Y_{Sloughed} = Y_{Obs.} = \frac{Y}{1 + bSRT_B} \tag{3.1}$$

Where:

$Y_{Sloughed}$	=	Yield of nitrifying organisms sloughed or detached into the bulk liquid relative to the amount of ammonia removed in the biofilm
Y	=	Growth yield of nitrifying organisms in the biofilm relative to the amount of ammonia removed. Also sometimes referred to as the "true" growth yield, mg COD/mg N
b	=	Decay rate of nitrifiers in the biofilm, d^{-1}
SRT_B	=	Retention time of nitrifiers in the biofilm, d

Although the model equation itself was not presented by the authors, model predictions were presented for over three years' data from the Duck Creek and Rowlett Creek WWTPs. Comparison with observed results showed the model could accurately predict effluent ammonia in the range of 1 mg N/L for much of the operating period as well; some, but not all, of the months showed higher effluent ammonia in the range of 5 mg N/L.

The inclusion of influent ammonia and nitrifying biomass in the model presented by Daigger *et al.* represented an important improvement in predictive power over the conventional design model from Equation 1.1.

3.5.2 SIDESTREAM/MAINSTREAM PROCESSES

Seeding of nitrifying organisms into the activated sludge process occurs in plants with sidestream nitrification or partial nitritation-anammox with wasting of sidestream biomass into the mainstream process. Plaza *et al.* (2001) evaluated the benefits of sidestream-to-mainstream seeding as part of a pilot study at the Uppsala WWTP [24]. As illustrated in Figure 3.6, this pilot included separate activated sludge trains to treat, respectively, high-strength dewatering reject water and primary effluent. The train treating primary effluent achieved effluent ammonia less than 1 mg

N/L during operational periods when SRT was less than 1.5 d and temperatures in the range of 13 to 16°C.

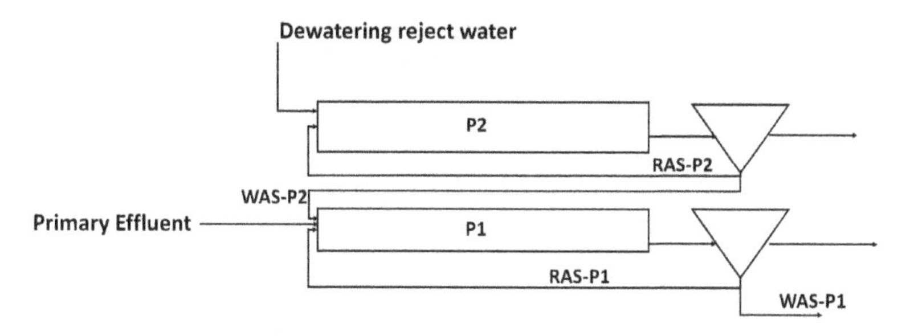

Figure 3.6 Process flow diagram for the sidestream treatment/mainstream treatment seeding strategy

The authors presented an analytical solution to the same mass balance problem described by Daigger *et al.*, namely to solve for effluent ammonia when influent nitrifying organisms are not negligible: $X_0 \neq 0$. The solution for the general case was developed treating the equations as a quadratic and solving for the roots as presented in Equation 3.2. This quadratic solution has the inconvenience of including a "±" which means that the (+) or (-) form of the equation needs to be used, depending on whether the operating SRT is above or below the washout SRT. The discontinuity introduced by this feature is evident from Figure 3.7.

To address the shortcomings of Equation 3.2, the authors also presented a simplified solution for the special case where $K_S = 0$. The simplified equation that results from this assumption does not address the fundamental problems of the quadratic solution approach, however. In Chapter 4, a more stable solution to the mass balance for a hybrid process will be developed. As opposed to the quadratic approach used by Plaza *et al.*, the basis for the equations in Chapter 4 is a "brute force"' approach using the symbolic math solution library available in the Python programming language.

$$S = \frac{A}{2} \pm \sqrt{\frac{A^2 - 4B}{4}} \tag{3.2}$$

Where:

A	=	$S_0 - \frac{K_N + \frac{X_0 \mu \tau}{Y}}{1 - \mu \tau}$
B	=	$-\frac{K_N S_0}{1 - \mu \tau}$
S_0	=	Influent ammonia concentration, mg N/L
X_0	=	Influent nitrifying biomass concentration, mg/L
S	=	Effluent ammonia concentration, mg N/L
K_N	=	Half saturation concentration for ammonia, mg N/L
μ	=	Nitrifier growth rate at given temperature, d^{-1}
Y	=	Growth yield of nitrifiers, mg/mg N
b	=	Decay rate of nitrifying organisms, d^{-1}
τ	=	Mixed Liquor Sludge Retention Time (SRT), d

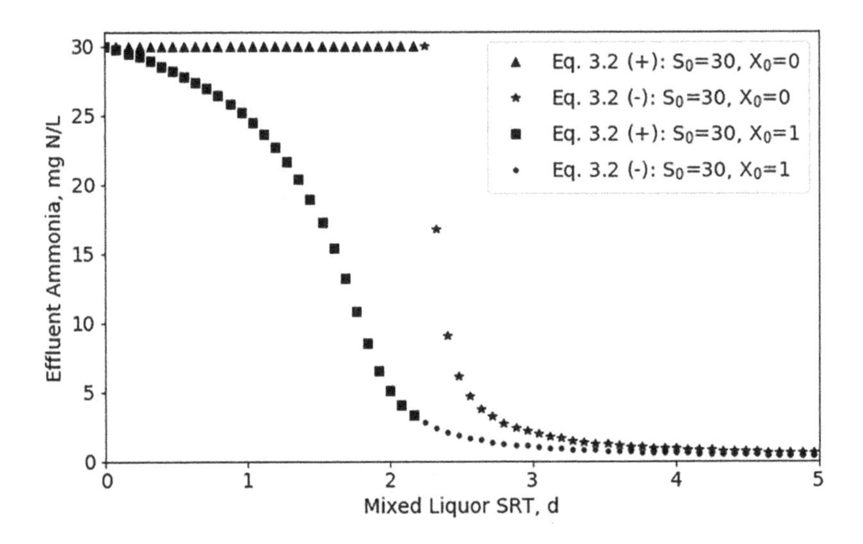

Figure 3.7 Nitrification washout curves assuming influent nitrifier seed of 0 and 0.25 mg/L

Plaza *et al.* do not discuss how influent nitrifying biomass should be quantified but concluded from model sensitivity analyses that seeding had the greatest benefit in processes that were operated below or near the critical (washout) SRT, and little benefit above that. It can be argued that Plaza *et al.* found little benefits at longer SRTs because they did not investigate system response to loading dynamics. The benefits of bioaugmentation, or seeding effect, for improving response to peak loading events will be explored in Chapters 4 and 6.

3.5.3 SEEDING PILOT

The design and operations of a "seeding pilot" to isolate the seeding effect of nitrifying organisms from an MABR biofilm on a downstream suspended growth chemostat reactor is described by Houweling *et al.* (2018) [14]. The system as presented in Figure 3.8 includes an unseeded reactor, Chemostat A, to act as a control to the "seeded" reactor, Chemostat B. Both reactor trains were fed at the same flow rate with a synthetic feed, based on a solution of ammonium bicarbonate and a nutrient broth, made up to an ammonia nitrogen concentration of 50 mg/L.

The MABR reactor contained 0.13 m^2 of hollow-fiber MABR media, resulting in an ammonia loading rate in a range of 5 to 25 $g/m^2/d$, depending on feed flow rate. The percent of the ammonia load removed in the MABR biofilm, $F_{Nit,B}$, was tuned through a combination of adjustments to influent loading and process airflow and averaged 30 to 50%.

Chemostats were selected to simulate the suspended growth in the activated sludge process because it allowed for precise control of SRT. There is, by definition, no difference between sludge retention time, SRT, and hydraulic retention time, HRT,

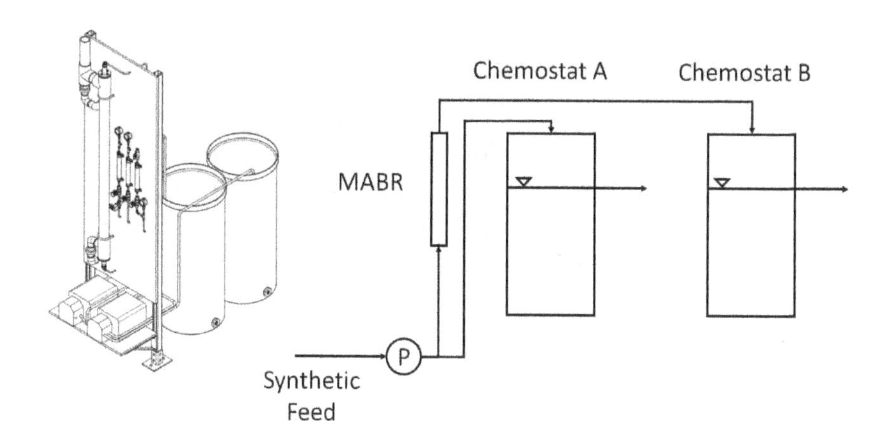

Figure 3.8 Isometric view (left) and process flow diagram (right) of "seeding pilot" showing MABR reactor mounted on panel with "unseeded" (Chemostat A) and seeded (Chemostat B)

Table 3.1
Operating Parameters for Seeding Pilot

Parameter	Units	Chemostat A	MABR	Chemostat B
Flow	mL/min	15 – 45	15 – 45	15 – 45
Volume	L	60	2	60
HRT	d	0.6 – 2.5	0.03 – 0.09	0.6 – 2.5
Bulk Dissolved Oxygen	mg/L	>4	0	>4
pH	S.U.	7 – 8	7 – 8	7 – 8
Media Specific Ammonia Loading Rate	$g/m^2/d$	N/A	5 – 25	N/A

in a chemostat reactor. Operating chemostat reactors served the purpose of eliminating questions around wasted activated sludge volumes or effluent suspended solids, which introduce significant uncertainty in estimates of SRT for conventional activated sludge pilots. In the seeding pilot, mixed liquor SRT was only a function of the feed flow and the chemostat volume. The former was set with a dual head peristaltic pump, which ensured that flows to Chemostat A and B were identical, and the latter was set by the standpipe height in the chemostat reactor tanks.

Reactor volumes and operating conditions are presented in Table 3.1, where the volume in the chemostat reactors was controlled using overflow standpipes internal to the reactors.

Results from August to November 2017 are presented in Figure 3.9 showing operational periods when the SRT of Chemostats A and B were progressively lowered from 2.5 d to a minimum of 0.7 d before finally being increased up to 1.5 d. All operations were conducted at a temperature of approximately 20°C, and changes to SRT were made to both chemostats by adjusting the height of the reactor overflow standpipe.

Results from Figure 3.9 are summarized as a "washout curve" in Figure 3.10 where the average ammonia concentration for steady-state periods are plotted as

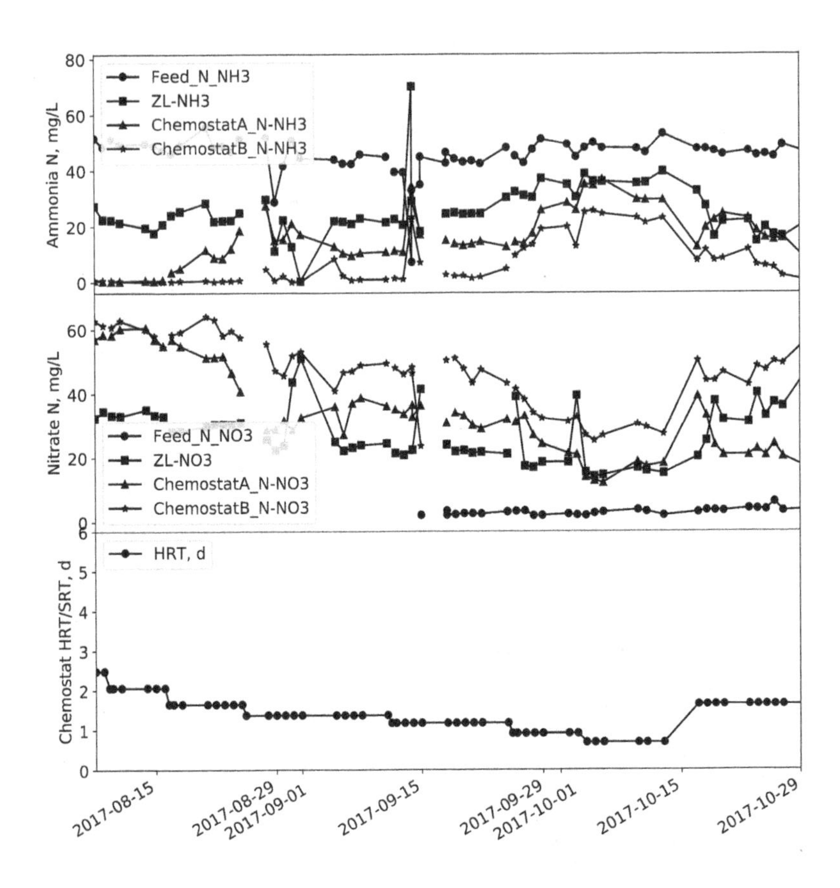

Figure 3.9 Results of "seeding pilot" operated at 20°C and varying chemostat HRT

a single point. These averaging periods were identified when the chemostats were judged to be operating at or near steady state, typically after waiting the equivalent of three SRTs following a change in chemostat standpipe height or other operational change.

As can be seen from Figure 3.10, the "seeded" chemostat achieved effluent ammonia less than 1 mg N/L at an SRT as low as 1.3 d. In comparison, the effluent from the "unseeded" Chemostat A achieved effluent ammonia less than 1 mg N/L only up to an SRT of 2.1 d. Below this threshold, effluent ammonia was greater than 10 mg N/L. The flattened washout curve of Chemostat B, as compared to Chemostat A, is evidence of the beneficial impact of bioaugmentation, or seeding effect, in this reactor.

It should be noted, however, that both chemostats showed some signs of "wall effect" nitrification despite receiving regular cleaning to remove biofilm growth. This is most clearly seen in the washout curve of Chemostat A, which didn't benefit from any seeding effect, but still never achieved complete washout of nitrifying organisms.

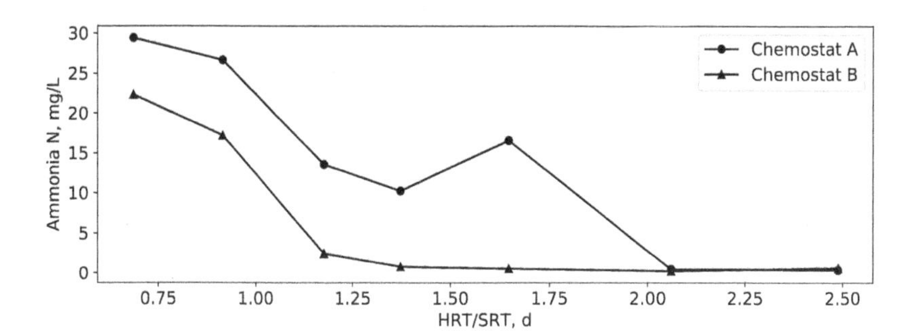

Figure 3.10 "Washout curves" at 20°C presented for Chemostat A and Chemostat B based on average results for steady-state periods

Equation 1.1 predicts that, at an SRT less than 1.5 days, the effluent from Chemostat A should have been equal to the feed ammonia concentration of 50 mg N/L. But the highest effluent ammonia concentration measured was 30 mg N/L, even at an SRT of 0.7 days. The explanation for this was likely the wall effect nitrification that was taking place in both chemostats.

The experimental setup described here provides a simple means to decouple and study the impact of biofilm on suspended growth behavior. Operation of an unseeded, control reactor provides a baseline for evaluating the hybrid process performance. And comparison of the seeded and unseeded washout curves provides an intuitive means for understanding the benefits of the hybrid process. Lessons learned from this pilot relate to the potential for "wall effect" nitrification which can be significant at low loading rates to the chemostat. Even with regular scrubbing to remove biofilm in the chemostats, nitrification rates on the order of 1 g/d per m^2 of wall surface are possible, and experimental design should account for this possibility.

4 Design Equations for Hybrid Systems

Discussion of hybrid processes in Chapters 3 and 4 evoked the "seeding effect" in hybrid processes, enabling suspended growth nitrification near or below the washout SRT. In addition, the beneficial impacts of nitrification in the biofilm for reducing the load of ammonia to be treated in the mixed liquor was discussed. The basis for quantifying these two effects was identified within the framework of the nitrifier mass balance presented in Chapter 1. The only difference being that, for hybrid systems, the assumptions from Chapter 1 of (a) no impact from influent ammonia concentration, and (b) a negligible influent concentration of nitrifying organisms, were no longer valid.

This theoretical framework is further explored in this chapter to develop tools to quantify performance of hybrid systems under a range of operating conditions. These tools can be used to parse out the relative contribution of ammonia removal in the biofilm, and the beneficial impact of "seeding effect" on mixed liquor nitrification.

More specifically, this chapter presents design equations to account for the following:

- Washout curves showing effluent ammonia as a function of mixed liquor SRT for different ranges of nitrification in the biofilm, $F_{Nit,B}$, and sloughing yield, $Y_{Sloughed}$.
- How increasing nitrification in the biofilm, $F_{Nit,B}$, enables operation at lower mixed liquor SRTs.
- How hybrid systems improve safety factor by increasing the nitrification potential of the mixed liquor, relative to the ammonia load it needs to remove, $R_{Nit,max/L}$.

4.1 REVISITING THE ACTIVATED SLUDGE WASHOUT CURVE

4.1.1 IGNORING INFLUENT BIOMASS

The Activated Sludge Process

Figure 4.1 presents the process flow diagram for a completely-mixed activated sludge (CMAS) process. From the point of view of mass balancing, CMAS represents the activated sludge process in its simplest form and, for this reason, is the basis of the equations in this chapter. Most activated sludge bioreactors are designed to include some degree of plug-flow behavior because it results in improved process behavior. But accounting for this would make derivation of design equations impossible. Notwithstanding, the equations that follow are still useful when applied to plug-flow processes and, indeed, they see widespread use in practice.

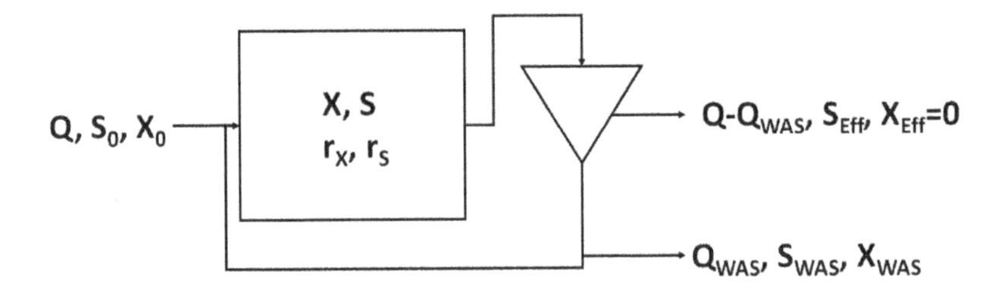

Figure 4.1 Process flow diagram of activated sludge process used to derive Equation 1.1

The symbols used in Figure 4.1 follow the convention where "S" represents biomass growth susbtrate and "X" represents biomass. For the purposes of this book, the substrate is ammonia and the biomass is nitrifying, ammonia oxidizing bacteria (AOBs). These symbols are defined as follows:

Q:	influent flow, m^3/d
Q_{WAS}:	waste activated sludge flow, m^3/d
S_0:	ammonia concentration in the influent, mg/L
S_{Eff}:	ammonia concentration in the effluent, mg/L
S_{WAS}:	ammonia concentration in the wastage stream, mg/L
S:	ammonia concentration in the bioreactor, mg/L
X_0:	nitrifying biomass concentration in the influent, mg/L
X_{Eff}:	nitrifying biomass concentration in the effluent, mg/L
X_{WAS}:	nitrifying biomass concentration in the wastage stream, mg/L
X:	nitrifying biomass concentration in the bioreactor, mg/L
V:	reactive volume of bioreactor, L
r_X:	conversion rate of biomass in the bioreactor, mg/L/d
r_S:	conversion rate of ammonia in the bioreactor, mg/L/d

An analytical solution to the mass balance from Figure 4.1 is considered in the following sections for the case where $X_0 = 0$. Section 4.2 will then consider the case where $X_0 \neq 0$, that is to say, the case where there is no seeding or bioaugmentation of nitrifying organisms into the mixed liquor.

Mass Balance Equations

Because the clarifier is assumed to be essentially non-reactive, both S_{EFF} and S_{WAS} can be assumed equal to S. Also, clarifier effluent suspended solids are assumed to be negligible compared to the wastage rate $Q_{WAS}X_{WAS}$, so $Q_{Eff}X_{Eff}$ is assumed to be zero. Finally, it is assumed that influent biomass can be neglected, *i.e.* $X_0 = 0$.

Based on these assumptions, the mass balances for ammonia and nitrifying organisms are presented in Equations 4.1 and 4.2, respectively.

$$In - Out + Reaction = Accumulation$$

$$QS_0 - QS + r_S V = \frac{d}{dt} SV = 0 \tag{4.1}$$

$$-Q_{WAS} X_{WAS} + r_X V = \frac{d}{dt} XV = 0 \tag{4.2}$$

In this chapter, steady-state conditions will be assumed and so for both equations $\frac{d}{dt} = 0$. Dynamic conditions, where this assumption is not true, will be explored in Chapter 6.

An important relationship that can be derived from Equation 4.1 is between the rate of substrate utilization r_S and the mass balance of substrate across the system $(S_0 - S)$ as follows:

$$r_S = -\frac{Q}{V}(S_0 - S) \tag{4.3}$$

Equation 4.3 will be used below in developing the equation to isolate for X.

Kinetics and Stoichiometry

To define the reaction rate of the ammonia and nitrifying organisms, we must define the process kinetics and the stoichiometry. The process kinetics are defined by the growth r_{Growth} and decay rates r_{Decay} of nitrifying organisms as follows:

$$r_{Growth} = \mu X \frac{S}{K_S + S} \tag{4.4}$$

$$r_{Decay} = bX \tag{4.5}$$

The stoichiometric relationships between the process kinetics and conversion rates can then be defined for nitrifying organisms, X, and ammonia, S:

$$r_X = r_{Growth} - r_{Decay} \tag{4.6}$$

$$r_S = -\frac{r_{Growth}}{Y} \tag{4.7}$$

Note that, by definition, the conversion rate, r_S, is in a negative relationship to r_{Growth} because, as X is generated $(+\Delta X)$, the result is removal of S $(-\Delta S)$. Some texts avoid use of negative terms by relating r_{Growth} to the rate of substrate utilization, r_{SU} which is, by definition, a positive term. Either approach is valid as long as it is applied consistently in all equations. The resulting conversion rates for ammonia, S, and nitrifying organisms, X, become:

$$r_S = -\frac{\mu X}{Y} \frac{S}{K_S + S} \tag{4.8}$$

$$r_X = (\mu \frac{S}{K_S + S} - b)X \tag{4.9}$$

An important relationship between r_X and r_S can be derived from these equations that will prove useful in isolating X from Equation 4.2. From Equations 4.6 and 4.7 we can derive the following relationship:

$$r_X = -Yr_S - bX \tag{4.10}$$

Isolating for X

The traditional way to develop the performance equation for the activated sludge process is to first manipulate the mass balance Equation 4.2 to isolate for X. The equation for X can then be substituted into Equation 4.1, which is then solved for S.

The approach first requires rearranging Equation 4.2 as follows:

$$Q_{WAS}X_{WAS} = r_X V \tag{4.11}$$

Equation 4.10 is then substituted into Equation 4.11 to give:

$$Q_{WAS}X_{WAS} = (-Yr_S - bX)V \tag{4.12}$$

Dividing both sides of the equation by XV we get:

$$\frac{Q_{WAS}X_{WAS}}{XV} = -\frac{Yr_S}{X} - b \tag{4.13}$$

Since SRT is defined as the mass of sludge in the bioreactor, XV, divided by the mass of sludge leaving the system, $Q_{WAS}X_{WAS}$, we can use the definition of $SRT = (XV)/(Q_{WAS}X_{WAS})$ to give the following:

$$\frac{1}{SRT} = -\frac{Yr_S}{X} - b \tag{4.14}$$

Substituting into Equation 4.14 the definition for r_S provided in Equation 4.3 we get:

$$\frac{1}{SRT} = \frac{YQ/V(S_0 - S)}{X} - b \tag{4.15}$$

Equation 4.15 can now be solved to isolate for X as follows:

$$X = \frac{SRT}{V/Q}\left(\frac{Y(S_0 - S)}{1 + bSRT}\right) \tag{4.16}$$

Solving for S

Equations 4.16 and 4.8 can now be substituted into Equation 4.1 to give the following:

$$QS_0 - QS + \left(-\frac{\mu\frac{SRT}{V/Q}\left(\frac{Y(S_0-S)}{1+bSRT}\right)}{Y}\frac{S}{K_S + S}\right)V = 0 \tag{4.17}$$

From which S can be isolated to give:

$$S = \frac{K_S(1 + bSRT)}{SRT(\mu - b) - 1} \tag{4.18}$$

Equation 4.18 was presented in Chapter 1 as Equation 1.1. It serves as the basis of the "washout curves" presented in Figure 1.1 with the limitation that it does not account for influent ammonia or biomass concentration. This limitation will be addressed in Section 4.1.2.

Insensitive to HRT and Influent Ammonia

It may be surprising that Equations 4.18 and 1.1 are dependent on SRT and nitrifier kinetic parameters, but not influent flow or concentrations. Not only are influent ammonia and nitrifying biomass absent from these equations, but neither is there any reference to volume, V, flow, Q, or hydraulic residence time, HRT. What this means is that the residence time of a drop of water in the activated sludge process has no bearing on the level of treatment it receives. The HRT does appear, however, in Equation 4.16 which identifies the concentration of nitrifying organisms in the bioreactor as proportional to the term SRT/HRT.

This provides insight into how the activated sludge process works: the term SRT/HRT represents the decoupling of the sludge and hydraulic retention time. It is a measure of how intensive the activated sludge process is compared to a simpler chemostat-type process, such as a lagoon, where $SRT = HRT$. Unlike activated sludge, the level of treatment in a lagoon process is very much dependent on the HRT.

For a typical activated sludge process with SRT of 10 days and HRT of 6 hours, the ratio of SRT/HRT would be 40. That is to say that, the biomass concentration in an activated sludge bioreactor might be on the order of 40 times higher than what would be expected in a chemostat reactor. This is a relationship that will prove useful later in this chapter.

As will be shown in the following sections, when considering influent biomass, *i.e.* seeding effect, the bioreactor HRT still does not influence the nitrification performance. Including influent biomass in the mass balance equations does have the beneficial effect, however, of leading to design equations that account for the impact of influent ammonia concentration.

4.1.2 ACCOUNTING FOR INFLUENT BIOMASS

Bioaugmentation, or seeding effect, can occur as a result of biomass in the influent. This was the case for the TF/AS process discussed in Chapter 3, Section 3.5.1: sloughed nitrifying organisms from the trickling filter biofilm seeded the downstream activated sludge process and enabled nitrification at a level that would otherwise not be possible. Seeding can also occur as a result of detachment and sloughing from a biofilm located inside the bioreactor. This is the case for the hybrid IFAS processes which were discussed in Sections 3.2 and 3.3. From the point of view of mass balancing, it is easier to consider the case where the seeded biomass is in the bioreactor influent, and so this is the approach taken in the following sections. But the resulting design equations are equally valid for both cases.

The Chemostat Reactor

For the case where X_0 is assumed to be zero, the V/Q term easily dropped out of the mass balance equations when substituting Equations 4.16 and 4.8 into Equation 4.1. This cancelling of terms is no longer achieved for the case of $X_0 \neq 0$, which makes solving these equations using analytical methods more difficult.

To keep the mass balance equations manageable, a chemostat reactor will be considered as the basis for mass balance equations with $X_0 \neq 0$. This introduces the useful assumption that $SRT = HRT$. Since HRT has no bearing on effluent ammonia concentration, the equation for effluent ammonia would be the same regardless of whether one assumes a chemostat or completely-mixed activated sludge, CMAS. So basing the mass balances on chemostat or CMAS changes nothing for design equations of effluent ammonia concentration. This is not the case for the equation of biomass concentration in the bioreactor, however. As discussed for Equation 4.16, the nitrifying biomass concentration in a CMAS bioreactor is higher than the concentration in a chemostat by a factor of SRT/HRT.

The process flow diagram for the chemostat reactor is presented in Figure 4.2.

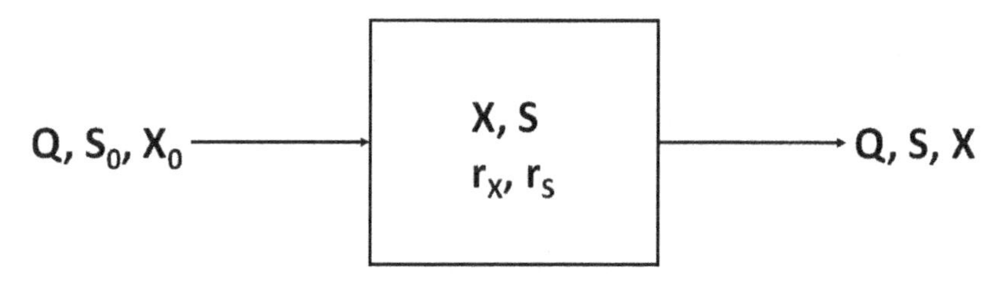

Figure 4.2 Mass balance for development of analytical solutions

The equivalence of design equations based on the CMAS process, presented in Figure 4.1, and the chemostat presented in Figure 4.2, will be demonstrated by benchmarking to Equations 4.16 and 4.18 for cases where $X_0 \neq 0$.

Mass Balance Equations

The mass balance equations for substrate S and biomass X in the chemostat reactor are presented in Equations 4.19 and 4.20:

$$In - Out + Reaction = Accumulation$$

$$QS_0 - QS + r_S V = \frac{d}{dt} SV = 0 \tag{4.19}$$

$$Q_0 X_0 - QX + r_X V = \frac{d}{dt} XV = 0 \tag{4.20}$$

Introducing HRT and SRT

The HRT of the chemostat reactor is the volume divided by the flow-rate :

$$HRT = \frac{V}{Q} \tag{4.21}$$

The SRT of the chemostat reactor is the mass of solids in the chemostat XV divided by the mass rate QX at which solids are lost in the effluent. Since X appears in both the numerator and denominator, the relationship for SRT is seen to be the same as for HRT:

$$SRT = \frac{XV}{QX} = \frac{V}{Q} \tag{4.22}$$

For expediency in developing further equations, the symbol τ will be substituted for V/Q instead of SRT or HRT:

$$\tau = HRT = \frac{V}{Q} \tag{4.23}$$

So using $\tau = V/Q$, and the definitions of the reaction terms r_X and r_S defined in Equations 4.8 and 4.9, we get the following equations:

$$\frac{S_0 - S}{\tau} = \frac{\mu X}{Y} \frac{S}{K_S + S} \tag{4.24}$$

$$\frac{X_0 - X}{\tau} = (\mu - b)X \frac{S}{K_S + S} \tag{4.25}$$

Approach for Developing Design Equations

The approach to developing the design equations for S and X in the following sections is based on isolating expressions for X from Equations 4.24 and 4.25. These expressions for X are not useful on their own because they still relate X to the unknown value of S. But the two expressions for X can be subtracted from each other to give an equation equal to zero. This equation is very useful indeed because it contains only S as an unknown variable. A symbolic math solver, such as SymPy, can then be used to find the roots to this equation.

SymPy is a symbolic math library that is freely distributed for the Python programming language. What is valuable about symbolic math programs is that they allow variables to be treated as symbols, instead of numbers. Design equations can thereby be derived wherein numerical evaluation of parameters can be deferred until the very end. The design equations presented below could not have been developed without it.

Isolating X from the Mass Balances on S

Isolating X from Equation 4.24 gives the following expression:

$$X = \frac{Y(K_S S_0 + S(-K_S - S + S_0))}{S\mu\tau} \tag{4.26}$$

Isolating X from the Mass Balances on X

Isolating X from Equation 4.25 gives the following expression:

$$X = \frac{X_0(K_S + S)}{(K_S b\tau + K_S + Sb\tau - S\mu\tau + S)} \tag{4.27}$$

Each of the variables X_0, S_0, K_S, b, μ, τ and Y in Equations 4.26 and 4.27 are ones that can be measured, set by the process operator, or otherwise assumed from literature parameters of nitrifying organisms. As such, only S and X are treated as unknowns.

Solving for S

A new equation is created by subtracting the expressions for X presented in Equations 4.26 and 4.27. An expression of "$X - X$" naturally is equal to zero:

$$\frac{Y(K_S S_0 + S(-K_S - S + S_0))}{(S\mu\tau)} - \frac{X_0(K_S + S)}{(K_S b\tau + K_S + Sb\tau - S\mu\tau + S)} = 0 \tag{4.28}$$

Equation 4.28 can then be solved for S using the "solve" function in the SymPy library. Note that the resulting Equation 4.29 was obtained using only the default settings in the "solve" function. It is possible that a more simplified solution could be obtained through adjustments to these settings.

$$S = \frac{A - \sqrt{B}}{C} \tag{4.29}$$

where:

$$A = K_S\tau Yb + K_S Y - S_0\tau Yb + S_0\tau Y\mu - S_0 Y + \tau X_0\mu$$

$$B = K_S^2\tau^2Y^2b^2 + 2K_S^2\tau Y^2b + K_S^2Y^2 + 2K_SS_0\tau^2Y^2b^2 - 2K_SS_0\tau^2Y^2b\mu + 4K_SS_0\tau Y^2b - 2K_SS_0\tau Y^2\mu + 2K_SS_0Y^2 + 2K_S\tau^2X_0Yb\mu + 2K_S\tau X_0Y\mu + S_0^2\tau^2Y^2b^2 - 2S_0^2\tau^2Y^2b\mu + S_0^2\tau^2Y^2\mu^2 + 2S_0^2\tau Y^2b - 2S_0^2\tau Y^2\mu + S_0^2Y^2 - 2S_0\tau^2X_0Yb\mu + 2S_0\tau^2X_0Y\mu^2 - 2S_0\tau X_0Y\mu + \tau^2X_0^2\mu^2$$

$$C = 2Y(\tau(\mu - b) - 1)$$

Equation 4.29 is rather complex and this is not unexpected since it is based on Equation 4.28, which is non-linear. It is virtually unrecognizable compared to the traditional design Equation 1.1 presented in Chapter 1. The denominator at least looks familiar when compared to the denominator from Equation 1.1 / 4.18: $(SRT(\mu - b) - 1)$, where $SRT = \tau$.

It should not be thought, however, that this equation is difficult to use in practice. Pasting this equation into a spreadsheet and referencing the parameters with "named cells" makes its use straightforward. An example of this is provided at the following website: "www.IntensifyingActivatedSludge.com" .

4.2 COMPARING THE TWO APPROACHES

4.2.1 PARAMETER VALUES

The results presented in Figure 4.3 are developed using the parameters presented in Table 4.1.

Table 4.1
Stoichiometric and Kinetic Parameters for Nitrifier Mass Balance

Parameter	Value	Units
$\hat{\mu}$	0.9	d^{-1}
b	0.17	d^{-1}
DO	2	mg O_2/L
K_{DO}	0.25	mg O_2/L
K_S	0.7	mg N/L
θ_μ	1.073	[-]
θ_b	1.029	[-]

And where the effect of dissolved oxygen limitation and temperature on the specific growth and decay rates are accounted for as follows:

$$\begin{aligned} \mu &= \hat{\mu}_{20C} \frac{DO}{K_{DO}+DO} \theta_\mu^{T-20} \\ b &= b_{20C} \theta_b^{T-20} \end{aligned} \tag{4.30}$$

The parameters in Table 4.1 have been selected because they reflect the values used in commonly used activated sludge models. The use of these models will be discussed in greater depth in the next chapter. Note that it is not appropriate to account for ammonia limitation in the above equation because it is already accounted for in the term $S/(K_S+S)$ we find in Equations 4.8 and 4.9.

4.2.2 WASHOUT CURVES

Although Equations 4.18 and 4.29 are based on different mass balances, they give identical predictions for the case of $X_0 = 0$. This is demonstrated in Figure 4.3 which shows exact overlap for the two. Equation 4.29, however, has the following advantages:

- The variable X_0 can be used to account for bioaugmentation, the so-called seeding effect. This is true whether the seeded nitrifying organisms be in the influent or sloughed from a biofilm located inside the bioreactor.

- The variable S_0 accounts for the effect of influent ammonia strength.
- Unlike Equation 4.18, Equation 4.29 presents no discontinuities at or below the critical SRT.

The value of not having discontinuities in Equation 4.29 is worth noting. Discontinuities can raise difficulties when outputs from one equation are being used as the input to other process calculations. This is very typical in spreadsheet based design tools. Because Equation 1.1 is discontinuous at the point of the washout SRT, it is more difficult to use in a design tool.

The power of Equation 4.29 will be further demonstrated in Chapter 5 by comparison with results generated using process simulation software.

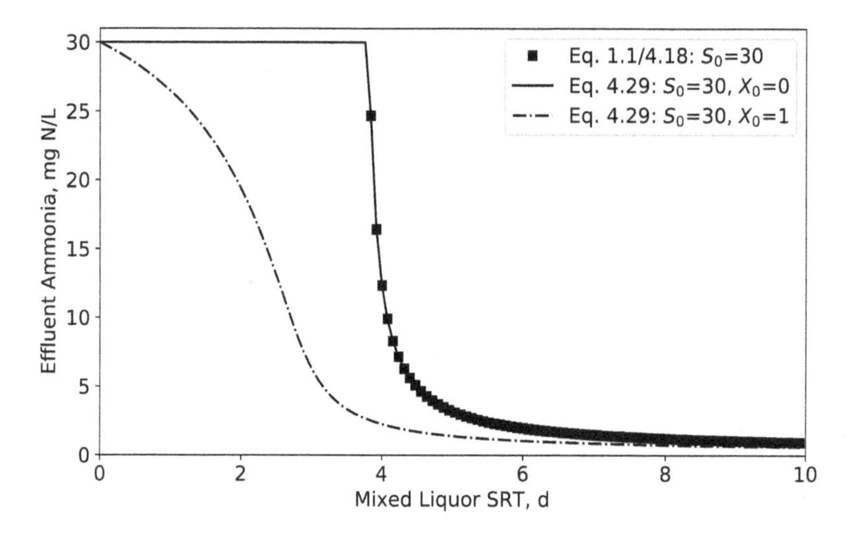

Figure 4.3 Washout curves as predicted by Equations 1.1/4.18 and 4.29

4.3 DESIGN EQUATIONS

The curves presented in Figure 4.3 are useful for demonstrating the equivalence between Equations 4.18 and 4.29. But they do not reflect the fact that nitrifier seeding comes as a result of ammonia removal in the biofilm. So more seeding, that is to say a higher X_0, should be accompanied by less ammonia for the mixed liquor to remove, *i.e.* lower S_0.

To illustrate, for the TF/AS process described in Section 3.5.1, the influent to the activated sludge stage, S_0, would be equivalent to the ammonia concentration in the plant influent, S_{Inf}, less the fraction of ammonia removed in the trickling filter biofilm, $F_{Nit,B}$. In the same way, for the IFAS processes described in Sections 3.2 and 3.3, the ammonia load on the mixed liquor would be proportional to the influent

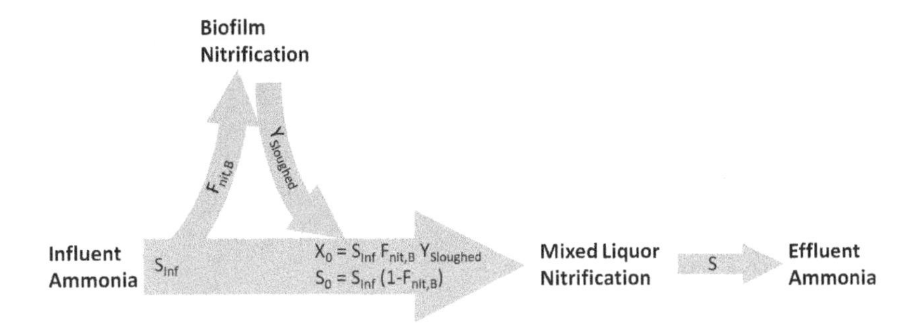

Figure 4.4 Illustration of relationship between S_{Inf}, $F_{Nit,B}$, $Y_{Sloughed}$, S_0 and X_0

ammonia concentration, S_{Inf}, and the fraction of the ammonia load removed in the biofilm, $F_{Nit,B}$.

The relationship between the fraction of the ammonia load removed in the biofilm, $F_{Nit,B}$, and the ammonia concentrations, S_{Inf} and S_0, is presented in Equation 4.31. The influent ammonia concentration, S_{Inf}, represents what needs to be treated by the hybrid system as a whole, whether it be TF/AS, IFAS or other. The ammonia concentration, S_0, represents what needs to be treated by the mixed liquor after removal of a fraction, $F_{Nit,B}$, in the biofilm.

$$S_0 = S_{Inf}(1 - F_{Nit,B}) \tag{4.31}$$

And the nitrifiers seeded to the mixed liquor X_0 can be defined based on the fraction $F_{Nit,B}$, and the sloughed yield $Y_{Sloughed}$ as follows:

$$X_0 = F_{Nit,B} S_{Inf} Y_{Sloughed} \tag{4.32}$$

The relationship between S_{Inf}, $F_{Nit,B}$, $Y_{Sloughed}$, S_0 and X_0 is illustrated in Figure 4.4.

Using Equations 4.31 and 4.32 to define S_0 and X_0, Equation 4.29 can be applied to compare designs with varying degrees of ammonia removal in the biofilm $F_{Nit,B}$, and assuming different yields of sloughed nitrifiers $Y_{Sloughed}$.

4.3.1 WASHOUT CURVES IN HYBRID PROCESSES

The usefulness of Equation 4.29 is not so much for predicting exact outcomes as it is for identifying relationships and trends. The washout curves presented in Figures 4.5, 4.6, 4.7, 4.8, 4.9 and 4.10 provide examples of how this might be done. These figures account for the impact on predicted effluent ammonia concentration of:

- Fraction of influent ammonia load removed in the biofilm $F_{Nit,B}$,
- Yield of nitrifying organisms sloughed from the biofilm, $Y_{Sloughed}$,
- Liquid temperatures of 10°C and 20°C, and
- Strength of influent ammonia concentration, S_{Inf}.

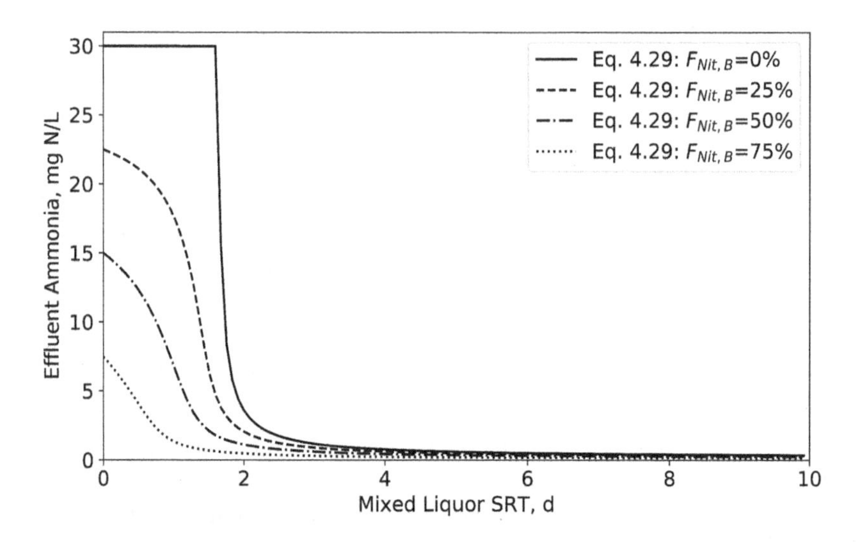

Figure 4.5 Washout curves developed for varying $F_{Nit,B}$: assuming $S_{Inf} = 30mgN/L$, 20°C and seeding effect, *i.e.* $Y_{Sloughed} = 0.05$

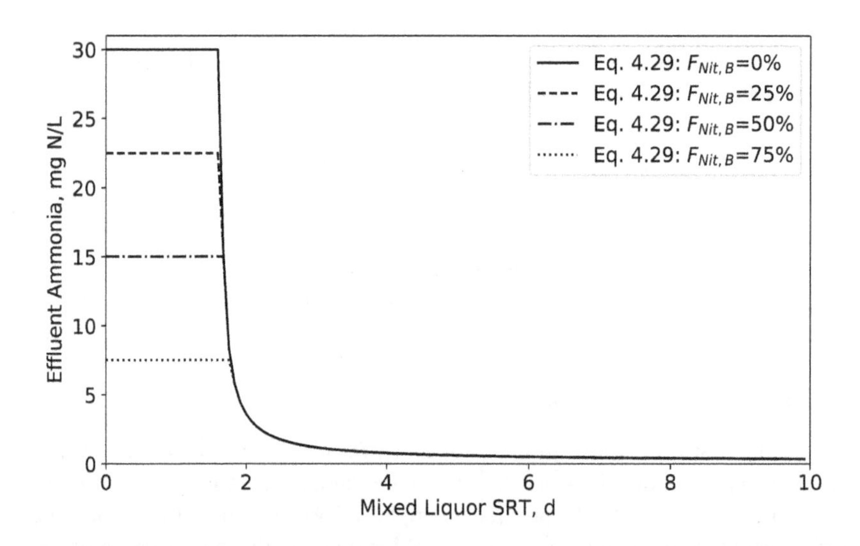

Figure 4.6 Washout curves developed for varying $F_{Nit,B}$: assuming $S_{Inf} = 30mgN/L$, 20°C, and no seeding effect, *i.e.* $Y_{Sloughed} = 0$

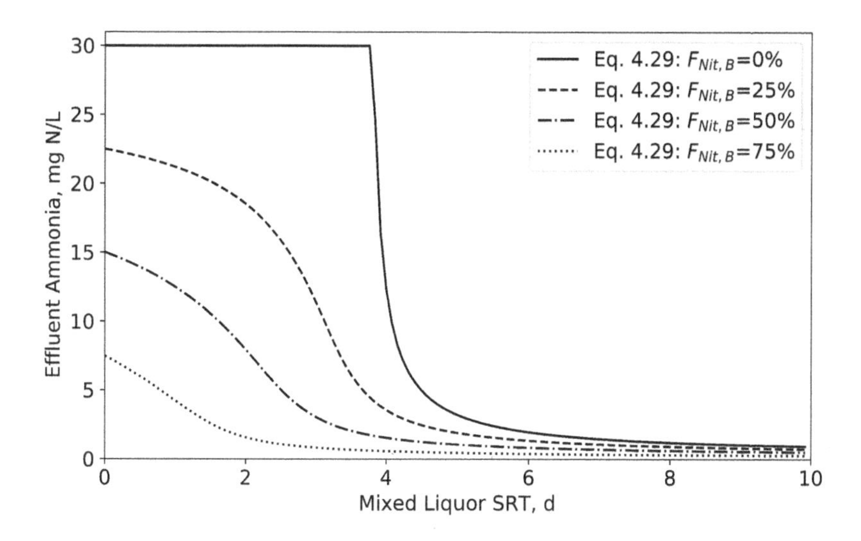

Figure 4.7 Washout curves developed for varying $F_{Nit,B}$: assuming $S_{Inf} = 30mgN/L$, 10°C and seeding effect, *i.e.* $Y_{Sloughed} = 0.05$

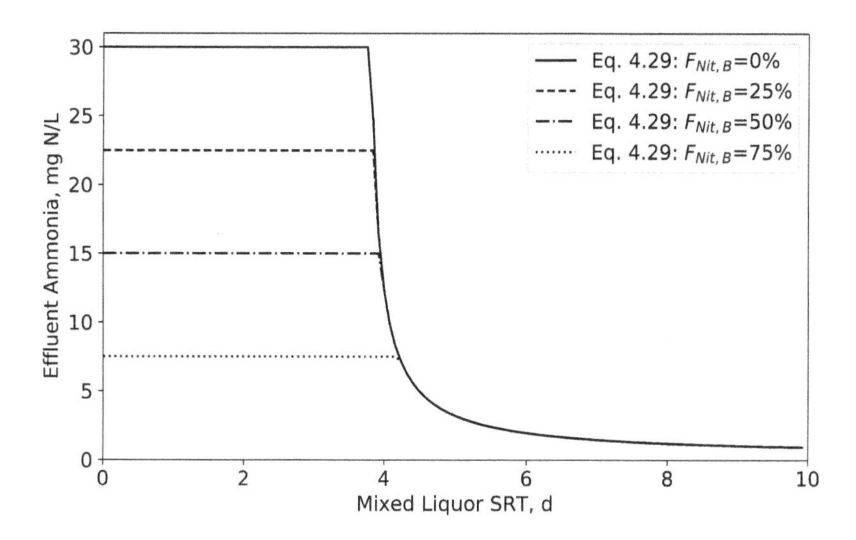

Figure 4.8 Washout curves developed for varying $F_{Nit,B}$: assuming $S_{Inf} = 30mgN/L$, 10°C, and no seeding effect, *i.e.* $Y_{Sloughed} = 0$

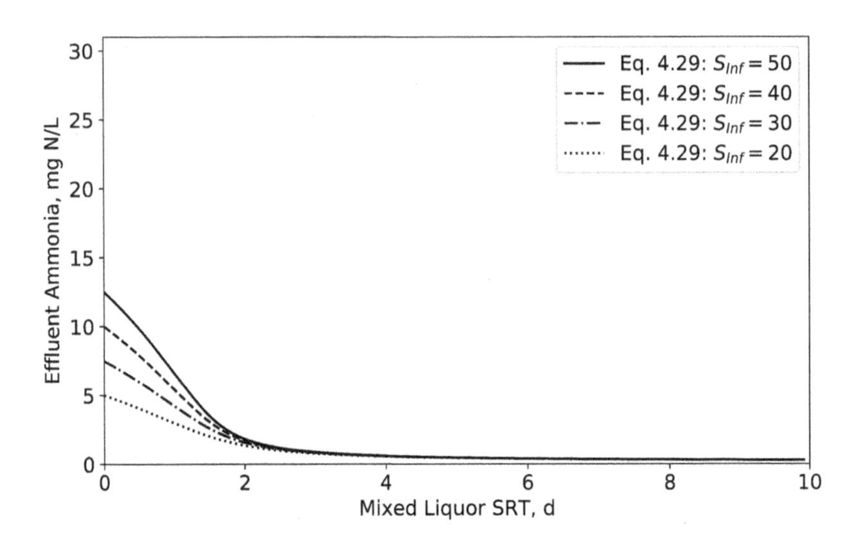

Figure 4.9 Sensitivity to influent ammonia concentration S_{Inf} assuming removal in the biofilm, $F_{Nit,B} = 75\%$, sloughing yield, $Y_{Sloughed} = 0.05$, and a temperature of 10°C

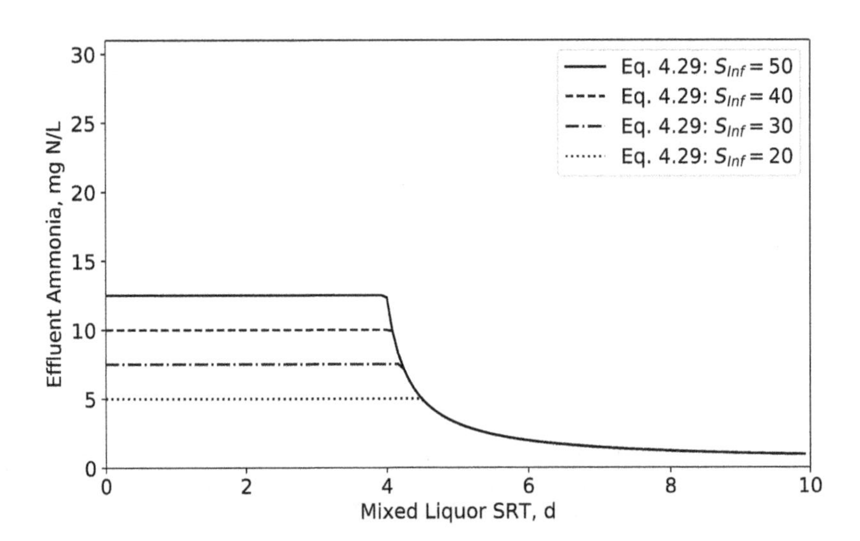

Figure 4.10 Sensitivity to influent ammonia concentration S_{Inf} assuming removal in the biofilm, $F_{Nit,B} = 75\%$, sloughing yield, $Y_{Sloughed} = 0$, and a temperature of 10°C

One immediate lesson from these curves is that the difference in effluent ammonia between conventional activated sludge ($F_{Nit,B}$=0) and hybrid processes ($F_{Nit,B}$ >0) is greatest at low SRTs. This suggests that the value of hybrid processes is greatest at low SRTs. However, this is a simplistic view of things. Even a small improvement in treatment efficiency, or reliability, can be extremely valuable if it means the difference between a plant meeting, or not meeting, its effluent treatment objectives. The value of hybrid processes operating at longer SRTs will be further explored in Section 4.3.4, as well as in Chapter 6.

The role of seeding effect can also be identified from these curves by comparing washout curves with $Y_{Sloughed}$ of 0 and 0.05, where $Y_{Sloughed}$ = 0 is used to simulate complete absence of seeding effect. Based on these comparisons, we see that seeding effect enhances the benefit of the hybrid process both below, at, and even above the washout SRT. This appears to be greatest right at the washout SRT but, as discussed, the benefit of hybrid processes needs to be understood in light of the effluent treatment objectives.

The Nitrifier Sloughing Yield, $Y_{Sloughed}$

As discussed in Chapter 3, Section 3.5.1, the yield of nitrifying organisms sloughed from the biofilm, $Y_{Sloughed}$, relates to their growth yield, Y, their retention time in the biofilm, SRT_B, and their rate of decay, b. The growth yield, Y, can be thought of as an "intrinsic" parameter. In principle, it can be derived from the stoichiometry of the individual catabolic and anabolic steps that enable nitrifying organisms to use ammonia, carbon dioxide, and other raw constituents as the basis for their growth.

The sloughing yield, $Y_{Sloughed}$, is analogous to the observed yield of the nitrifiers in the biofilm. While Y only accounts for the anabolic energy requirements of biomass growth, $Y_{Sloughed}$ additionally accounts for the maintenance energy required to keep the organism metabolically active after it has grown. In addition, $Y_{Sloughed}$ accounts for the decay of nitrifiers in the biofilm that may result from other processes including lysis and predation. The effects of maintenance, lysis, predation and any other process that results in "decay" of nitrifiers are lumped together in a single decay parameter, b.

$$Y_{Sloughed} = \frac{Y}{1 + bSRT_B} \tag{3.1}$$

Where:

$Y_{Sloughed}$	:	Yield of nitrifying organisms sloughed or detached into the bulk liquid relative to the amount of ammonia removed in the biofilm
Y	:	Growth yield of nitrifying organisms in the biofilm relative to the amount of ammonia removed. Also sometimes referred to as the "true" growth yield, mg/mg N
b	:	Decay rate of nitrifiers in the biofilm, d^{-1}
SRT_B	:	Retention of nitrifiers in the biofilm, d

The benefits of bioaugmentation, or seeding effect, in hybrid processes are intuitive. Mass balancing dictates that if nitrifiers continue growing in the biofilm, and

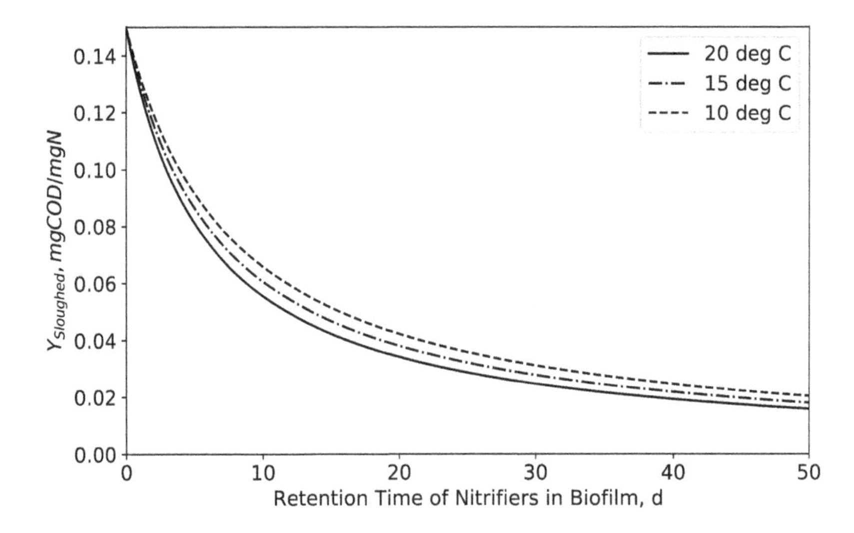

Figure 4.11 Relationship between nitrifier sloughing yield and retention time in the biofilm

biofilm thickness is not changing, then sloughing or detachment must be occurring. But to what extent? The parameter $Y_{Sloughed}$ is useful because it defines the seeding effect in terms of the mass of nitrifiers seeded to the suspended mixed liquor, and the mass of ammonia nitrified in the biofilm. As presented in Equation 3.1, the magnitude of $Y_{Sloughed}$ depends on the retention time of nitrifying organisms in the biofilm, SRT_B. While SRT_B itself cannot be measured directly, it relates to biofilm thickness, which can be measured. Modeling indicates that a range of 6 to 30 days is reasonable to assume when biofilm thickness is 500 μm or thinner. Using Equation 3.1, the relationship between $Y_{Sloughed}$ and biofilm retention time, SRT_B, is presented in Figure 4.11.

Theoretical Washout

The theoretical washout SRT can be identified in Figures 4.6, 4.8 and 4.10 at the point of inflection of the horizontal lines. Below the washout SRT there is no nitrification at all in the mixed liquor. (There is no obvious washout SRT in Figures 4.5, 4.7 and 4.9 due to the effect of seeding effect, as accounted for by $Y_{Sloughed} = 0.05$.) Washout SRTs of just below 2 and approximately 4 days are identified from these figures for temperatures of 20 and 10°C, respectively. Alternately, washout SRTs can also be calculated based on the inverse of the nitrifying organisms net growth rate:

$$SRT_{Washout} = (\mu - b)^{-1} \tag{4.33}$$

Where Equation 4.30 is used to relate μ to dissolved oxygen and temperature, we get

$$SRT_{Washout} = 0.9d^{-1}\frac{2mg/L}{0.25mg/L+2mg/L}1.072^{(20-20)} - 0.17d^{-1}$$

$$SRT_{Washout} = 1.6d$$

We can also compare the results from Figure 4.5, which indicate a washout SRT of around 4 days, with the theoretical washout SRT when the temperature is 10°C:

$$SRT_{Washout} = 0.9d^{-1}\frac{2mg/L}{0.25mg/L+2mg/L}1.072^{(10-20)} - 0.17d^{-1}$$

$$SRT_{Washout} = 4.4d$$

4.3.2 MINIMUM SRT CURVES

An alternative way to use Equation 4.28 is to treat effluent ammonia S as the fixed, or target, variable and SRT or τ as the variable to be solved for. Solving Equation 4.28 for τ gives Equation 4.34. Equation 4.34 is somewhat simpler than Equation 4.29 even though both are derived from Equation 4.28. The solution for τ, however, shows the same hallmark signs of non-linearity as the solution for S.

$$\tau = -Y\frac{K_SS - K_SS_0 + S^2 - SS_0}{K_SSYb - K_SS_0Yb + S^2Yb - S^2Y\mu - SS_0Yb + SS_0Y\mu + SX_0\mu} \quad (4.34)$$

Where:

$$S_0 = S_{Inf}(1 - F_{Nit,B})$$

and:

$$X_0 = F_{Nit,B}S_{Inf}Y_{Sloughed}$$

Equation 4.34 is a simpler expression than Equation 4.29 because it does not need to solve the roots to the S^2 term that appears when Equation 4.28 is expanded. More importantly, however, is that by relating $F_{Nit,B}$ to a reduction in the required SRT, Equation 4.34 allows the benefits of the hybrid system to be evaluated more directly. Recall from Chapter 1 that there is a direct relationship between SRT and the required bioreactor and clarifier tank volumes. So by identifying the "SRT savings" that a hybrid solution provides, Equation 4.34 can be used to identify capital cost savings from deferred construction of bioreactor tanks and clarifiers.

Although Equation 4.34 conveys the same information as Equation 4.29, it presents discontinuities at some values of its independent variable, S. In contrast, Equation 4.29 is continuous for all positive values of its independent variable τ. The reason for this is because there are effluent values of S that cannot be achieved through variation of τ alone.

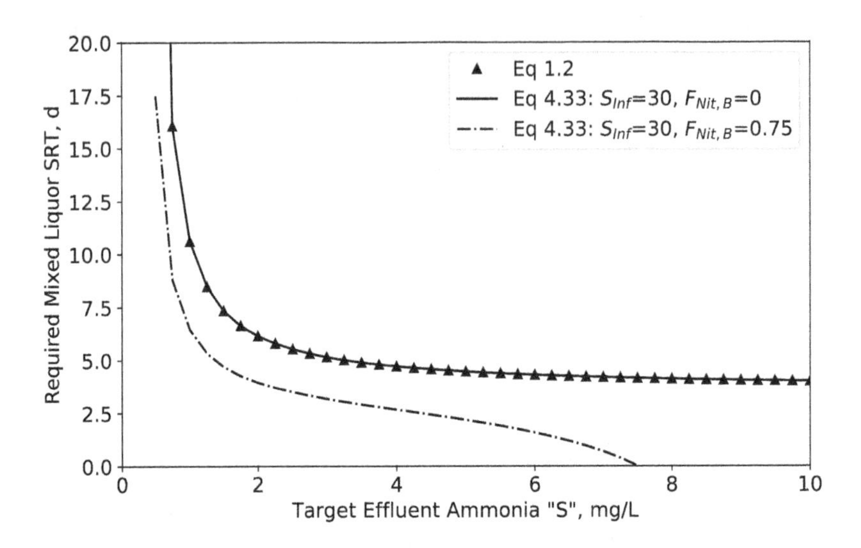

Figure 4.12 Relationship between required SRT and effluent ammonia objective for $F_{Nit,B}$ =0 and 75% assuming nitrifier sloughed yield $Y_{Sloughed}$ = 0.05

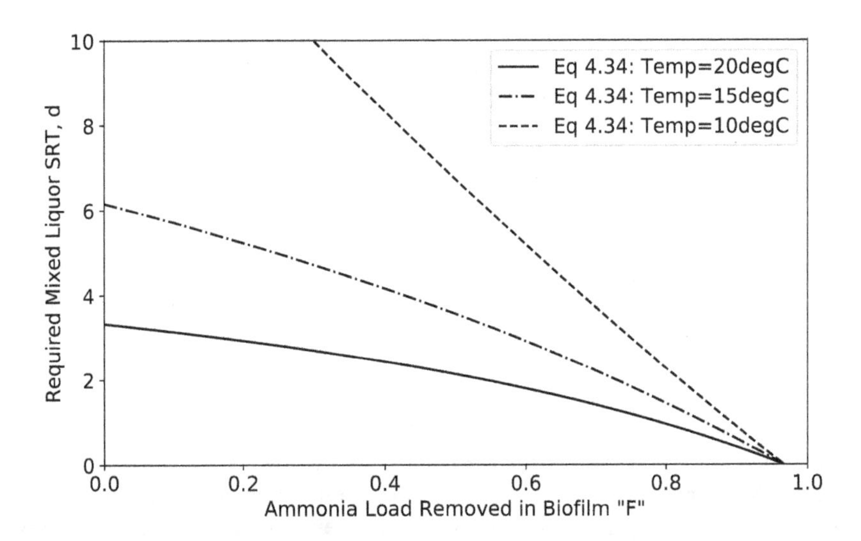

Figure 4.13 Relationship between required SRT and fraction on influent ammonia $F_{Nit,B}$ assuming nitrifier sloughed yield $Y_{Sloughed}$ = 0.05

To validate Equation 4.34, curves based on the traditional equation for SRT, presented in Chapter 1, Equation 1.2, are presented in Figure 4.12. Curves from Equations 1.2 and 4.34 match perfectly for the conventional activated sludge case where $F_{Nit,B} = 0$.

Figure 4.12 suggests that hybrid processes can provide the greatest savings in SRT, and therefore capital cost, when the effluent ammonia objective is higher. As discussed in Chapter 1, higher effluent ammonia objectives are common for plants whose effluent objectives are based primarily on toxicity, oxygen demand and/or where the effluent receiving water has a high dilution capacity.

Since hybrid processes tend to be designed based on some target ammonia removal in the biofilm, $F_{Nit,B}$, it is more helpful to use Equation 4.34 to develop curves of required mixed liquor SRT as a function of $F_{Nit,B}$. Figure 4.13 presents these curves which reveal a steep, linear slope at 10°C, with declining slopes at 15 and 20°C. This indicates that hybrid processes provide greater benefit when mixed liquor nitrification is stressed, in this case due to colder temperatures.

Equation 4.34 is a powerful tool for generating rapid insights into the effect of temperature, influent ammonia concentration, S_{Inf}, fraction of the ammonia load removed in the biofilm, $F_{Nit,B}$, and seeding effect, $Y_{Sloughed}$, on require mixed liquor SRT. Commercial simulation software could also have been used to generate the same insights, but it would take a lot longer. In fact design equations and simulation software are complementary tools. Whereas the design equations presented in this chapter are valid only for simple configurations, simulation software can help extend the relationships from Figure 4.12 to any type of process configuration.

4.3.3 REQUIRED AMMONIA REMOVAL ON BIOFILM

The relationship between required SRT, fraction of ammonia load removed in the biofilm, $F_{Nit,B}$, and the target effluent ammonia, S, identified in Equation 4.34 can be reorganized to isolate $F_{Nit,B}$. This may be a preferable way to regard this relationship when answering the question: how much ammonia removal in the biofilm, $F_{Nit,B}$, is required to achieve my effluent objective given my SRT constraint?

The solution for $F_{Nit,B}$ is presented in Equation 4.35:

$$F_{Nit,B} = \frac{-Y(K_S(S\tau b + S - \tau S_{Inf} b - S_{Inf}) + S(S\tau b - S\tau\mu + S - \tau S_{Inf} b + \tau S_{Inf}\mu - S_{Inf}))}{(S_{Inf}(K_S\tau Y b + K_S Y + S\tau Y b - S\tau Y\mu + S\tau Y_{Sloughed}\mu + SY))} \tag{4.35}$$

Equation 4.35 is the basis for Figures 4.14, 4.15, 4.16 and 4.17 which identify the ammonia removal, $F_{Nit,B}$, required to meet the effluent target at a given SRT. Figures 4.14 and 4.15 assume higher seeding effect, $Y_{Sloughed}$=0.05, whereas Figures 4.16 and 4.17 assume lower seeding effect, $Y_{Sloughed}$=0.01. Note that it is not possible to assume no seeding effect in these curves, $Y_{Sloughed}$=0, because Equation 4.35 is undefined for this condition.

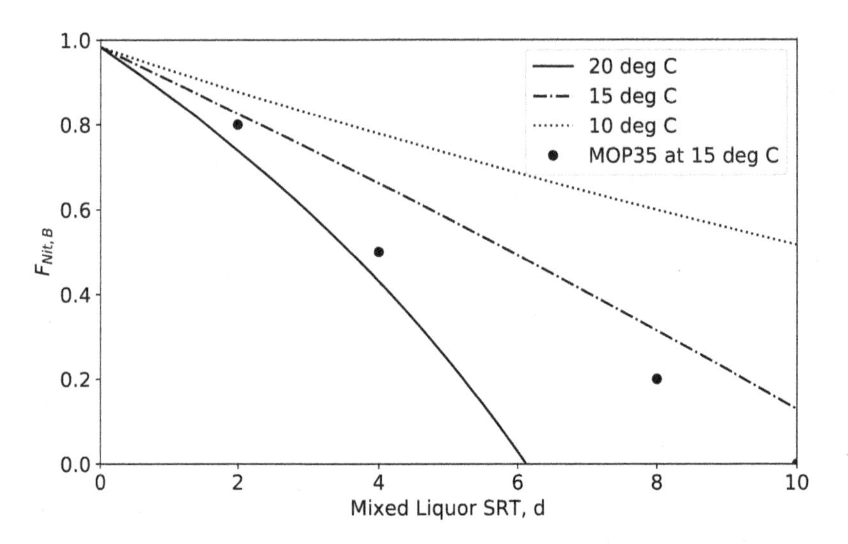

Figure 4.14 Required fraction of ammonia load to be removed on biofilm $F_{Nit,B}$ to meet an effluent ammonia of 0.5 mg N/L assuming seeding consistent with $Y_{Sloughed}$=0.05

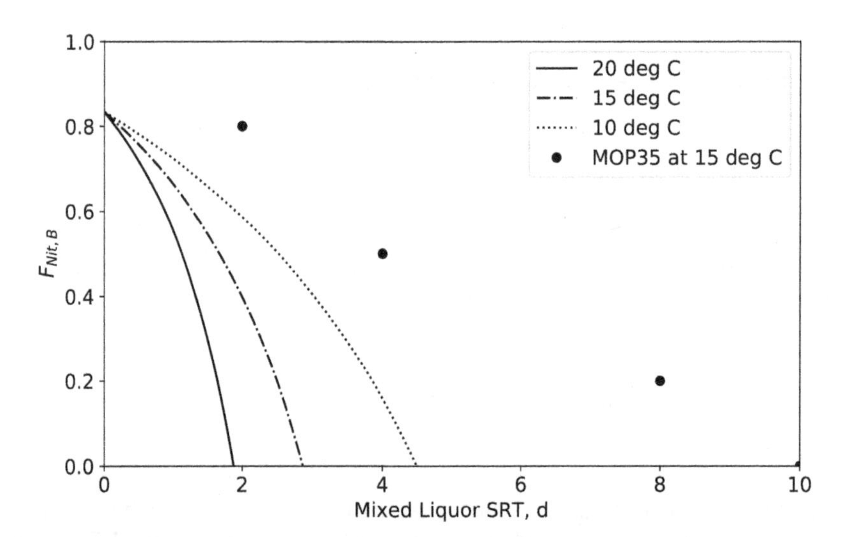

Figure 4.15 Required fraction of ammonia load to be removed on biofilm $F_{Nit,B}$ to meet an effluent ammonia of 5 mg N/L assuming seeding consistent with $Y_{Sloughed}$=0.05

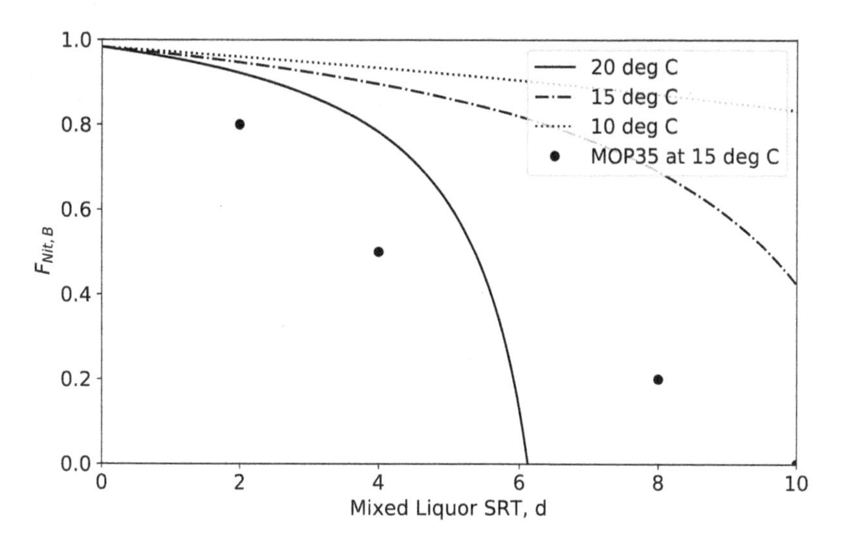

Figure 4.16 Required fraction of ammonia load to be removed on biofilm $F_{Nit,B}$ to meet an effluent ammonia of 0.5 mg N/L assuming seeding consistent with $Y_{Sloughed}$=0.01

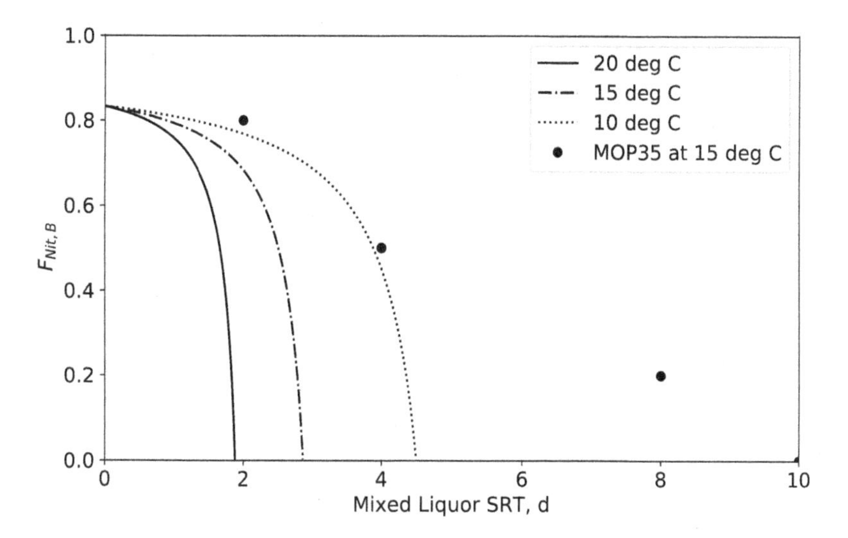

Figure 4.17 Required fraction of ammonia load to be removed on biofilm $F_{Nit,B}$ to meet an effluent ammonia of 5 mg N/L assuming seeding consistent with $Y_{Sloughed}$=0.01

Benchmarking to Design Guidelines

The curves in Figures 4.14, 4.15, 4.16 and 4.17 can be benchmarked to design guidelines from the WEF Manual of Practice on Biofilm Reactors, MOP 35, which state the following for 15°C [9]:

- At a 2-day mixed liquor SRT, 50% COD and 80% nitrification is on the biofilm. The rest is in the mixed liquor.
- At a 4-day mixed liquor SRT, 25% COD and 50% nitrification is on the biofilm. The rest is in the mixed liquor.
- At an 8-day mixed liquor SRT, 20% nitrification is on the biofilm. The rest is in the mixed liquor.

The closest match between Equation 4.35 and the MOP 35 design guidelines is achieved in Figure 4.14 which assumes an effluent ammonia target of 0.5 mg N/L and a seeding effect proportional to $Y_{Sloughed}$=0.05. The fact that, in Figure 4.14, the "MOP 35" points fall below the curve for 15°C indicates that it may be appropriate to assume an even stronger seeding effect, *i.e.* higher $Y_{Sloughed}$. Note that, unlike the MOP 35 design guidelines, Equation 4.35 does not link COD removal and nitrification in the biofilm. It is assumed in this text that nitrification is the limiting process in hybrid system design and that $F_{Nit,B}$ should be the basis of design. In addition, predicting the ratio of COD removal to nitrification in the biofilm is outside the scope of Equation 4.35: it varies based on process configuration and whether the type of biofilm being used is co-diffusional or counter-diffusional. A process designer would have to carefully consider these factors when selecting the quantity of biofilm support media needed to achieve the design $F_{Nit,B}$.

Besides benchmarking the seeding effect to MOP 35 design guidelines, Equation 4.35 can be used to identify the sensitivity of hybrid system design to effluent ammonia objectives. For example, there is a notable difference in shape of the curves when designing for an effluent ammonia objective of 0.5 mg N/L, in Figures 4.14 and 4.16, as compared to 5 mg N/L, in Figures 4.15 and 4.17. Very steep curves in Figures 4.15 and 4.17 reflect the fact that only a small amount of biofilm nitrification is required to help a conventional activated sludge plant meet an effluent objective of 5 mg N/L at or near the washout SRT. But as SRTs decrease below this, the requirement for nitrification on the biofilm increases very quickly. Designing for an effluent ammonia objective of 5 mg N/L can have important consequences for process reliability, particularly in cases where control of SRT is likely to be unreliable.

In comparison, Figures 4.14 and 4.16 show that when the effluent ammonia objective is only 0.5 mg N/L, a much greater proportion of the work needs to be done on the biofilm, and over a wide range of SRTs. But the process is likely to be much more stable than had it been designed for an effluent objective of 5 mg N/L.

4.3.4 HYBRID PROCESSES FOR SAFETY FACTOR ENHANCEMENT

Evaluating the benefits of hybrid processes based on washout curves presented in Figures 4.5 through 4.10 indicates much greater benefit when operating at or below the washout SRT. Also, the benefits seem to be greater when targeting higher effluent ammonia concentrations. Does this mean hybrid processes provide limited value when operated at long SRTs and low effluent objectives?

To answer this question, we can revisit the topic of safety factor first discussed in Chapter 1. Conventional design relies on extended SRTs to achieve the desired safety factor. The result is often design SRTs two or three times longer than would be dictated by initial evaluation of the washout curve presented in Figure 1.1. The rationale for this was illustrated in Chapter 1, Figure 1.2, wherein operating at longer SRTs multiplies the inventory of nitrifying organisms. A higher inventory of nitrifiers means a higher maximum potential nitrification rate as a ratio to the ammonia load to be treated, $R_{Nit,max/L}$. This ratio can also be calculated for hybrid processes by developing equations for the inventory of nitrifying organisms in the bioreactor, M_{AOB}, the maximum potential nitrification activity of this mass, $A_{Nit,max}$, and the effective load of ammonia that needs to be treated by the mixed liquor, L_{ML}.

As in Chapter 1, Equation 1.4, the concentration of nitrifiers in the biofilm can be calculated based on the removal of ammonia by the mixed liquor, ΔS, and accounting for the intensification factor SRT/HRT. For a hybrid system, however, the concentration of nitrifiers also needs to be accounted for in terms of the nitrifiers seeded through biofilm sloughing, X_0:

$$X_{AOB} = \frac{SRT}{HRT}\frac{Y\Delta S + X_0}{1 + bSRT} \tag{4.36}$$

where:

$$\Delta S = S_{Inf}(1 - F_{Nit,B}) - S$$

$$X_0 = F_{Nit,B}S_{Inf}Y_{Sloughed}$$

and recalling the definition for effluent ammonia, S, provided in Equation 4.29. The mass of nitrifying organisms in the mixed liquor is therefore simply the concentration multiplied by the bioreactor volume, V:

$$M_{AOB} = X_{AOB} \times V \tag{4.37}$$

The maximum potential activity of this mass of nitrifiers, M_{AOB}, is calculated based on the specific activity, which is their specific growth rate, μ, divided by their growth yield, Y:

$$A_{Nit,max} = X_{AOB} \times \frac{\mu}{Y} \tag{4.38}$$

And recalling that the ammonia load that the mixed liquor must treat is proportional to the fraction removed in the biofilm:

$$L_{ML} = QS_{Inf}(1 - F_{Nit,B}) \tag{4.39}$$

The maximum potential nitrification rate as a ratio to the ammonia load to be treated, $R_{Nit,max/L}$, can thus be calculated by dividing the maximum potential activity

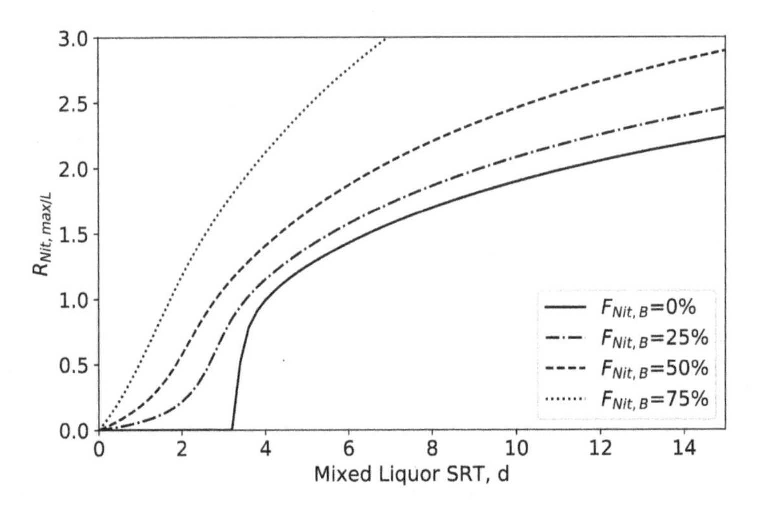

Figure 4.18 Ratio of nitrifying activity to effective load on the mixed liquor for a temperature of 10°C

of the mixed liquor, $A_{Nit,max}$, by the influent load to be treated in the mixed liquor, L_{ML}:

$$R_{Nit,max/L} = \frac{A_{Nit,max}}{L_{ML}} \tag{4.40}$$

By relating the maximum potential nitrification activity of the mixed liquor to the average load it must treat, Equation 4.40 represents the ability of the mixed liquor to treat peak loads. That is to say, loads over and above the average loads that the biomass is accustomed to seeing. This is equivalent to the ratio, $R_{Nit,max/L}$, presented in Chapter 1, Figure 1.2, which was developed based on results from process modeling software.

Design curves based on Equation 4.40 are presented in Figure 4.18 to demonstrate how $R_{Nit,max/L}$ increases both as a function of mixed liquor SRT, and the fraction of the ammonia load removed on the biofilm, $F_{Nit,B}$. To interpret these curves, we can begin by evaluating the conventional case where there is no biofilm, $F_{Nit,B} = 0$. In this case, $R_{Nit,max/L}$ is equal to 1 right around the washout SRT which, at 10°C, is just below 4 days. What this means is that there are sufficient nitrifying organisms to treat the average influent load at a 4 day SRT, but no more. There is no ability to treat any peaks in influent ammonia load. The ability to treat peaks in load is provided in conventional design by applying a safety factor to the SRT. Doubling the SRT from 4 to 8 days results in an increase of $R_{Nit,max/L}$ from approximately 1 to 1.8. In comparison, the same increase in $R_{Nit,max/L}$ from 1 to 1.8 can be achieved at an SRT of 4 days by increasing the ammonia removal in the biofilm from 0% to somewhere between 50 and 75%. How do we explain this? As illustrated in Figure 4.4, hybrid processes do two things:

1. By removing a portion of the ammonia in the biofilm, the effective ammonia load on the mixed liquor is decreased.
2. Just like in conventional activated sludge, the inventory of nitrifying organisms in the mixed liquor grows proportionally to the effective load it receives, and its SRT. However, hybrid processes also benefit from bioaugmentation, or seeding, of nitrifying organisms grown in the biofilm that subsequently "slough off" or detach into the mixed liquor.

Figure 4.18 presents how the ability to treat peak loads, as quantified by $R_{Nit,max/L}$, increases with respect to two parameters: mixed liquor SRT and the fraction of ammonia removal in the biofilm, $F_{Nit,B}$. Since managing peak loading events is one of the key reasons for application of safety factor, Figure 4.18 demonstrates how increasing $F_{Nit,B}$ can be used as an alternative to increasing SRT in hybrid systems.

4.4 HOW DOES A HYBRID PROCESS CREATE CAPACITY?

The preceding sections present the various design equations that can be derived for hybrid activated sludge processes. The hybrid process mass balance is distinguished from conventional activated sludge in that (a) a fraction of the ammonia load, $F_{Nit,B}$, is removed in the biofilm, and (b) there is a yield, $Y_{Sloughed}$, of nitrifying organisms grown in the biofilm that slough off, or detach, into the mixed liquor. The design equations provide insight into how hybrid systems:

- Shift and flatten the nitrification washout curve: Figures 4.5 to 4.8,
- Respond to differences in influent ammonia strengths: Figures 4.9 to 4.10.
- Decrease the required mixed liquor SRT to achieve a target effluent ammonia objective: Figures 4.12 and 4.13.
- In accordance with MOP 35 design guidelines, enable operation at decreased mixed liquor SRT: Figures 4.14 to 4.17.
- Provide an alternate means for managing peak loading events, *i.e.* achieving safety factor: Figure 4.18.

A hybrid process creates capacity by reducing the SRT required to nitrify in the mixed liquor. At SRTs near or below the washout SRT, this enables nitrification under conditions that would otherwise not be possible. At SRTs above washout, the main benefit is to increase the inventory of nitrifying organisms relative to the average influent ammonia load. This increases the ability of the mixed liquor to treat peak loading events. In either case, hybrid processes enable operation at a lower SRT than would otherwise be possible. Because bioreactor tank and secondary clarifier sizes are directly tied to mixed liquor SRT and influent load, hybrid processes provide a means to increase the effective capacity of conventional activated sludge.

4.5 SUMMARY

The design equations derived in this chapter are presented in Table 4.2. Design equations were validated by comparison with conventional design equations, for the case where $F_{Nit,B} = 0\%$, and with MOP 35 design guidelines for $F_{Nit,B}$ of 20%, 50% and 80% [9]. Further validation is presented in Chapter 5 by comparing results from Equation 4.29 with steady-state washout curves generated for hybrid processes using process simulation software. Dynamic simulations are explored in Chapter 6 to support the use of Equation 4.40 to quantify the ability of hybrid processes to improve management of peak loading events.

Typical biokinetic parameters to be used with these equations are presented in Table 4.1. Example implementations of these equations can be found at www.IntensifyingActivatedSludge.com.

Table 4.2
Design Equations

To calculate the effluent ammonia concentration:

Equation 4.29 $S = \frac{A-\sqrt{B}}{C}$

where:

$A = K_S \tau Y b + K_S Y - S_0 \tau Y b + S_0 \tau Y \mu - S_0 Y + \tau X_0 \mu$

$B = K_S^2 \tau^2 Y^2 b^2 + 2K_S^2 \tau Y^2 b + K_S^2 Y^2 + 2K_S S_0 \tau^2 Y^2 b^2 - 2K_S S_0 \tau^2 Y^2 b\mu + 4K_S S_0 \tau Y^2 b - 2K_S S_0 \tau Y^2 \mu + 2K_S S_0 Y^2 + 2K_S \tau^2 X_0 Y b\mu + 2K_S \tau X_0 Y \mu + S_0^2 \tau^2 Y^2 b^2 - 2S_0^2 \tau^2 Y^2 b\mu + S_0^2 \tau^2 Y^2 \mu^2 + 2S_0^2 \tau Y^2 b - 2S_0^2 \tau Y^2 \mu + S_0^2 Y^2 - 2S_0 \tau^2 X_0 Y b\mu + 2S_0 \tau^2 X_0 Y \mu^2 - 2S_0 \tau X_0 Y \mu + \tau^2 X_0^2 \mu^2$

$C = 2Y(\tau(\mu - b) - 1)$

To calculate the nitrifier growth rate:

Equation 4.30 $\mu = \hat{\mu} \frac{DO}{K_{DO}+DO} \theta^{T-20}$

To calculate the effective ammonia load on the mixed liquor:

Equation 4.31 $S_0 = S_{Inf}(1 - F_{Nit,B})$

To calculate the nitrifier seeding:

Equation 4.32 $X_0 = F_{Nit,B} S_{Inf} Y_{Sloughed}$

To calculate the yield of nitrifiers sloughed from the biofilm:

Equation 3.1 $Y_{Sloughed} = \frac{Y}{1+bSRT_B}$

To calculate the required SRT to meet effluent ammonia objective S:

Equation 4.34 $\tau = -Y \frac{K_S S - K_S S_0 + S^2 - S S_0}{K_S S Y b - K_S S_0 Y b + S^2 Y b - S^2 Y \mu - S S_0 Y b + S S_0 Y \mu + S X_0 \mu}$

To calculate the required fraction of ammonia load to remove on biofilm:

Equation 4.35 $F_{Nit,B} = \frac{-Y(K_S(S\tau b + S - \tau S_{Inf} b - S_{Inf}) + S(S\tau b - S\tau\mu + S - \tau S_{Inf} b + \tau S_{Inf}\mu - S_{Inf}))}{(S_{Inf}(K_S \tau Y b + K_S Y + S\tau Y b - S\tau Y \mu + S\tau Y_{Sloughed}\mu + SY))}$

To calculate the concentration of nitrifying organisms in the bioreactor:

Equation 4.36 $X_{AOB} = \frac{SRT}{HRT} \frac{Y\Delta S + X_0}{1+bSRT}$

To calculate the ammonia removal in the mixed liquor:

$\Delta S = S_{Inf}(1 - F_{Nit,B}) - S$

To calculate the mass of nitrifying organisms in the mixed liquor:

Equation 4.37 $M_{AOB} = X_{AOB} \times V$

To calculate the maximum potential nitrifying activity of the mixed liquor:

Equation 4.38 $A_{Nit,max} = X_{AOB} \times \frac{\mu}{Y}$

To calculate the effective ammonia load on the mixed liquor:

Equation 4.39 $L_{ML} = Q S_{Inf}(1 - F_{Nit,B})$

To calculate the ratio of the maximum potential nitrifying activity to effective ammonia load:

Equation 4.40 $R_{Nit,max/L} = \frac{A_{Nit,max}}{L_{ML}}$

---

5 Biofilm Modeling Using Simulation Software

The design equations presented in Chapter 4 are useful for assessing sensitivity of hybrid systems to a range of conditions. However, their power is limited by the simplicity of their underlying assumptions: steady-state conditions and a single-tank CSTR hydraulic regime. Most cases will significantly diverge from these assumptions. So the design equations from Chapter 4 should primarily be seen as indicative of relationships and sensitivities, rather than exact outcomes. To account for the complexities of plug-flow conditions, internal recycles, unaerated zones and plant dynamics, simulation software is required.

The purpose of this chapter is to present an overview of how hybrid biofilm systems are modeled using process simulation software. The behavior these models predict well is highlighted as well as some of the challenges they face. More specifically, the following items are discussed:

- 1-d models as the common basis for biofilm models in the commercially available software packages.
- the basic equivalence, for simple cases, of 1-d model predictions and the design equations from Chapter 4,
- the most useful predictions from 1-d models being the conditions required to establish biofilm nitrification, the effects of sloughed nitrifiers on the mixed liquor, and dynamic response to peak loading events, and
- indirect modeling of sloughing through a combination of surface detachment and an "internal solids exchange".

5.1 BIOFILM MODELING APPROACHES

5.1.1 HISTORY OF BIOFILM MODELING

Horn and Lackner noted in their "Modeling of Biofilm Systems: A Review" that the challenge of modeling of biofilms, as compared to suspended growth, relates to substrate gradients [13]. It can be added that, for hybrid systems, an equally important challenge is quantifying the interactions between the biofilm and the mixed liquor. While the wasting of mixed liquor, and SRT, can be measured and quantified without too much difficulty, measuring the sloughing and/or detachment rates from a biofilm to the mixed liquor is much more challenging. In practice, biofilm thicknesses are controlled by application of shear forces from mixing or aeration energy. But any experimental quantification of sloughing or detachment rates is likely to be approximate at best: the control is certainly not as direct or as precise as it is for waste activated sludge pumping.

According to Horn and Lackner, biofilm modeling began with simple one-dimensional (1-d) models in the 1970s [13]. The quality of these simulation results was later improved by consideration of mass transfer at the bulk/biofilm interface and detachment of biomass from the surface. By the 1990s, 1-d biofilm modeling was maturing through use of multispecies models that accounted for stratification of autotrophic nitrifying organisms and heterotrophs between the base and outside of the biofilm [35]. This markedly improved the predictive power of biofilm models for simulating biofilm flux rates and long-term behavior.

The move to two- (2-d) and three-dimensional (3-d) models in the 1990s was motivated by the desire for more realistic predictions of the surface structure of biofilms. In addition, these models began to include extracellular polymeric substances (EPS) in the biofilm structure. Recent research has focussed on developing models that calculate shear forces, with the benefit reportedly being more powerful predictions of surface detachment and 3-d structure. In addition, the elastic properties of biofilms are being explored in terms of how this impacts their resistance to shear. Overall, the efforts of researchers has focussed primarily on increasing the descriptive power of biofilm models. However, improvements to the predictive power for engineered systems remain, for the most part, unproven. The simple reason for this is that these models have not been adopted by practitioners.

Current activity among the research community engaged in biofilm model development is mostly disconnected from practice. This is unfortunate because practice has much to inform researchers about which biofilm behaviors are most relevant to process behavior. From a practitioner's point of view, the recent activity of biofilm researchers might be summarized as follows:

Increased model complexity

↓

Increased descriptive power

↓

Impossible to validate predictive power for engineering applications

5.1.2 THE CURRENT STATE-OF-THE-ART

The activity and progress in biofilm modeling over the last 20 years is undeniable, however, commercial process simulators remain firmly entrenched in the 1-d approaches developed in the 1990s. As a result, only 1-d models are being "field tested". As noted by Takács *et al.* (2007), this is because the amount of computer time required to solve 2-d models, never mind 3-d models, would be impractical for everyday engineering use [32]. Indeed, even 1-d models can be very numerically intensive, in many cases taking ten times longer to simulate than comparable suspended growth models.

But even if the issue of simulation time can be overcome, it may also be the case that 2-d and 3-d models offer little added value compared to simpler 1-d models currently in use. Greater model complexity does not necessarily translate to greater predictive power. State-of-the-art 3-d biofilm models may provide a lot of added resolution when looking at biofilm structures, but they are still far from capturing the heterogeneity and complexity of biofilms in full-scale activated sludge basins.

If one thinks, for example, about detachment and sloughing of a biofilm in a hybrid process, it is reasonable to question whether a 3-d model could meaningfully improve on the predictive power of a 1-d model. Even if the 3-d model could accurately describe the factors and mechanisms that govern these processes, the input data required to simulate the model would likely be unobtainable. This is discussed in further detail in the following paragraphs.

BOD Loading

BOD loading promotes the growth of hetetrophic organisms which can outcompete nitrifiers for available oxygen and space in the biofilm. In addition, they are likely to contribute to thicker biofilms resulting in higher diffusional resistance into the inner layers of the biofilm where nitrifiers tend to grow. At very high heterotrophic growth rates, the relative amounts of EPS and other slimy material generated in the biofilm may increase. It has been suggested that this type of slimy material is more likely to resist scour, contribute to high biofilm thicknesses and even media "bridging".

Influent BOD loading, on its own, can be highly variable and good characterization difficult to achieve. But the BOD loading on the biofilm is also dependent on the activity of the surrounding mixed liquor. How much of the influent BOD is removed or adsorbed in the bulk mixed liquor? Do byproducts of mixed liquor flocs, whether they be particulate or soluble, attach to or otherwise impact the biofilm composition? State-of-the-art "ASM" biokinetic models struggle to adequately address these questions and, as a result, the predictive power of hybrid process models is impacted.

Mixing

One of the hallmarks of process modeling is the use of completely-mixed compartments, or series of completely-mixed compartments, to represent movement of bulk liquid between bioreactor tanks, biofilm layers or other zones of interest. This is an effective modeling strategy when non-idealities in flow, potential dead zones, or, in the case of biofilms, the uneven distribution of scour energy across the biofilm surface do not govern process behavior. When this is not the case, computational fluid dynamics (CFD) and related techniques may provide added value. CFD models use first principles to describe important physical behavior but accounting for the interactions between multiple phases (liquid, solids and gas bubbles) can be an intractable problem, to say the least. These models still rely on empirical relationships to describe behavior at interfaces, and empirical relationships require data for calibration.

Observations of biofilms from full-scale systems often reveal patterns at both the microscopic and macroscopic scale. The biofilm is simply not uniform across the entire media surface area, and yet that is how 1-d models treat them. 1-d models only account for heterogeneity across the biofilm depth. 2-d or even 3-d models might prove more useful than 1-d models if they could simulate the influence of these patterns on process behavior. The data gathering required for model calibration and everyday use, however, might prove prohibitive.

Influent Wastewater Characteristics

Influent BOD and TSS are composed of particulate, dissolved and colloidal fractions that can be biodegradable or non-biodegradable, organic or inorganic. The relative fractions of these vary over the course of the day and week according to residential, commercial and industrial schedules. The influent dissolved biodegradable material, the so called readily biodegradable COD, is more likely to impact the behavior and oxygen utilization in the biofilm, but this really depends on the location of the biofilm in the process. Presence of trash and small particles, such as textile fibers, may adhere to the biofilm and cause a diffusion barrier.

Because of the way in which they impact biofilm behavior, influent wastewater characteristics are a critical input to any process model. However, very complex models will require more complex influent characterization models. In practice, the degree model of complexity that is warranted is limited by the availability of influent characterization data. Many plants only record influent BOD_5 and TSS data.

Higher Life Forms

Protozoa and nematode growth in biofilms can be abundant. The influence of these higher life forms on process kinetics is ignored in most design manuals, although there is anecdotal evidence suggesting that excessive "grazing" can de detrimental to nitrification rates. "ASM" type biokinetic models only explicitly account for the growth of bacteria and so, as a result, any other type of activity gets lumped in with heterotrophic bacteria during model calibration. The decay rate of heterotrophic bacteria in ASM models therefore includes the impact of protozoan grazing.

5.1.3 ROBUST MODELS FOR ROBUST DESIGNS

Typically, a robust design is one where scour energy can maintain an acceptable biofilm thickness in spite of the multitude of varying factors that are outside the operator's power to control, or the modeler's power to quantify. At best, an operator can hope to have feedback signals to know when the process needs adjustment. In this spirit, robust models do not need to describe every possible aspect of what goes on inside a biofilm, only the governing processes and mechanisms in a robust design.

When considering the robustness of a process model, it is helpful to keep in mind a list of things that the model predicts well, or well enough for the purposes of the modeling study, and a list of things the model predicts poorly or not at all. For hybrid processes there are, in fact, two models whose predictive power needs to be considered: the biokinetic model and the mass transport model. An "ASM" style biokinetic model is typically used to account for competition between nitrifiers and heterotrophic bacteria. The 1-dimensional biofilm model is the industry standard for modeling the structure and transport processes inside the biofilm. So then an example list, by no means exhaustive, of the strengths and weaknesses for "ASM" models is presented below.

ASM models predict:

Well or well enough	Poorly or not at all
– Sludge production – Oxygen requirements – Nitrification inventory as a proxy for nitrification activity – Nitrification response to dynamic loading variation	– NOB suppression – Hydrolysis kinetics – Kinetics of BOD removal

In a similar manner, a list for the 1-d biofilm model is presented below.

1-d biofilm models predict:

Well or well enough	Poorly or not at all
– Stratification perpendicular to the media surface – Surface detachment and erosion – Biofilm thickness, dry solids content and average biofilm SRT	– Variation parallel to the media surface – Sloughing – Biofilm SRT for individual components

These lists are provided only for the purpose of provoking thought and their content and descriptors could be the subject of endless discussion and debate. For example, one could question whether 1-d models really provide good predictions of biofilm thickness, or whether this needs to be calibrated on a case-by-case basis given the *in situ* mixing and scouring energy. The value of the process model would then be in calculating the biofilm SRT, and sloughing yield, $Y_{sloughed}$, given the user defined biofilm thickness and BOD loading.

5.1.4 PARSIMONIOUS MODELS

One of the central arguments in this chapter is that there is much to be gained from the use of simple models which account for only the most basic and relevant process behavior. This can be accomplished in a more direct and transparent way in a simple model than it can in a complex model. For example, the predictive power of a 1-d biofilm model can be shown to be very similar, for steady-state conditions, with the 2-compartment model shown in schematic form in Figure 5.1. This 2-compartment model is an attempt at model "parsimony": including the minimum number of mechanisms required to predict the desired behavior.

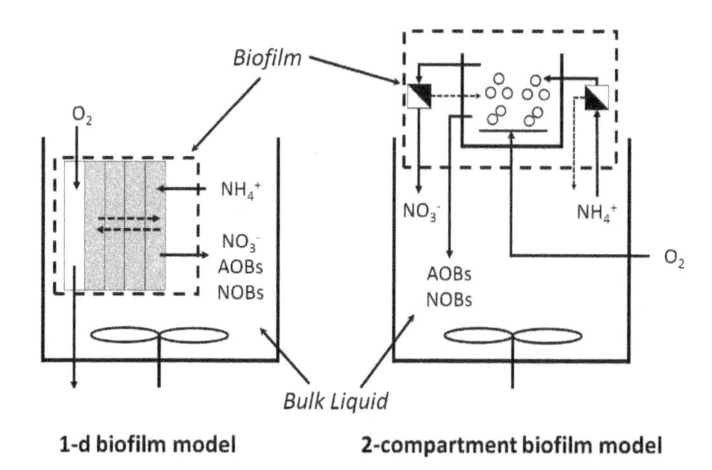

Figure 5.1 Schematic representation of 1-d and 2-compartment MABR biofilm model

Model Transparency

To calibrate the 2-compartment model, the user chooses the nitrification rate by adjusting the flow into the biofilm zone, and chooses the seeding effect, $Y_{Sloughed}$, by adjusting flow-rate of nitrifiers back from the biofilm to the bulk mixed liquor. This is illustrated in Figure 5.1. The 2-compartment model is not predictive of either of these biofilm properties and yet, once calibrated, it can perform very useful calculations of the impact of the biofilm on the bulk mixed liquor and overall process behavior.

Since nitrification rate and seeding effect are, in most cases, the two most critical biofilm properties in hybrid processes, they deserve particular attention. The 2-compartment model lets their magnitude be selected explicitly rather than have them be, as in a 1-d model, an emergent property of the multiple interactions taking place across the biofilm layers. This may be preferable when the model is being used for design. Nitrification rate, for example, is often a design basis parameter having been measured during pilot testing, or even guaranteed by a process equipment vendor. So a 2-compartment model that allows this rate to be fixed across all simulation runs is useful.

Although nitrification rate and seeding effect are emergent properties of 1-d biofilm models, identifying their magnitude can be very difficult. It usually involves performing mass balances across the biofilm reactor block, followed by trying to parse out the relative contributions to ammonia removal and nitrifier growth of the biofilm and mixed liquor. In most commercial simulation software it is a laborious process that likely takes place in a spreadsheet the user must set up. But without identifying the nitrification rate and seeding effect, there is no transparency with respect to the impacts of the biofilm on the overall hybrid process. With respect to this transparency, a 2-compartment model may be a superior option to a 1-d model in many cases.

Model Ease of Implementation

A 2-compartment model also has the advantage of being easily implemented within existing software packages for cases when either (a) no model block is available for the biofilm process being investigated or, as is often the case, (b) simulation of the flowsheet using the default 1-dimensional model is numerically too intensive and therefore too time consuming.

At the most basic level, a parsimonious model can be a simple look-up table or design curve that relates a flux rate to a process condition. Examples of these are the flux curves presented by Hem *et al.* and Rusten *et al.* [11, 26]. The goals and benefits of parsimonious models is described as follows:

Identify target behavior of biofilm

↓

Include minimum number of mechanisms
or relationships required to predict behavior

↓

Predictive capacity against full-scale systems readily assessed

The 1-d models included in most commercial simulation software are not parsimonious. They are descriptive, they are reasonably complex, and they can, in many cases, be extremely slow to simulate and difficult to calibrate. Nevertheless, engineers tolerate these difficulties because they value the added descriptive power that comes with 1-d model complexity. 1-d models help to understand what is going on inside the biofilm "black-box". Also, given the range of process scenarios that need to be evaluated, it is difficult to always know *a priori* which behavior the model needs to predict. So, one might argue that it is better to have a model that predicts as many behaviors as possible, and sort out which are important after the fact.

As will be shown later in this chapter, however, the use of 1-d models can in itself be dangerous. Careful attention must be paid to understanding the parameter calibration and how this impacts emergent behavior of the biofilm, notably the nitrification rate, biofilm thickness, and the yield of nitrifiers sloughed from the biofilm, $Y_{Sloughed}$.

5.2 SIMULATION SOFTWARE IMPLEMENTATIONS

5.2.1 1-DIMENSIONAL MODELS

The schematic view of a 1-d model presented in Figure 5.1 shows four compartments to represent four biofilm layers and one compartment to represent the bulk mixed liquor. Exchanges of soluble material between the bulk liquid and the biofilm layers is accounted for using diffusion rates based on Fick's Law. This is a sound mechanistic approach and is applied in essentially the same manner across all commercial process simulation packages.

Unlike movement of soluble components, the movement of microorganisms and other particulate material between the biofilm and bulk liquid in 1-d models does not have a sound mechanistic basis. In reality, sloughing can occur right from the base of the biofilm. So for that particular location in the biofilm, every layer detachs from the support media. Sloughing can occur in local areas of the biofilm without meaningfully impacting the biomass content of the entire biofilm. They are localized events.

In a 1-d model, however, there is no "local" area but rather a single uniform biofilm surface. So a sloughing event would apply to the biofilm in its entirety and this would be catastrophic to process performance: all of the biofilm would be lost at the same time. For this reason, only surface detachment is modeled in commercial 1-d biofilm models. In addition to being unable to model sloughing, this introduces the problem of how to refresh the contents of the inner layers of the biofilm that are not exposed to surface detachment. Without some mechanism to "refresh their contents", these layers would entirely fill up with the inert residue of biomass decay. Only the outer layer of the biofilm would then provide any kind of activity.

There are a variety of approaches used in commercial process model software to address this issue. These include use of "internal solids exchange rates" or "solids movement factors". Typically the impact of these parameters is poorly explained, if at all, in the available help documentation. A summary is provided as follows:

- A high internal exchange will result in less stratification of biofilm composition. This may be undesirable for biofilms that are expected to include both nitrifying and denitrifying layers.
- A high internal exchange rate will increase the rate at which organisms from inner layers reach the surface and are exposed to detachment. This may be desirable to simulate the effect of frequent sloughing events in which organisms from all layers are detached.
- A low internal exchange rate will result in stronger stratification between layers but may result in a lower rate of detachment of organisms from inner layers.

5.2.2 2-COMPARTMENT MODELS

For the 2-compartment model presented in Figure 5.1, model "parsimony" is achieved by focussing on the minimum number of processes to account for (a) the flux of ammonia into the biofilm, *i.e.* nitrification rate, and (b) the seeding effect, or rate of detachment, of nitrifying organisms returned from the biofilm to the mixed liquor, $Y_{Sloughed}$. A 2-compartment model may be a good alternative to a 1-d model when one of the following applies:

- The simulation platform does not have a model block for the process or technology of interest.
- Simulations using the 1-d biofilm model is too time-consuming. Experience has shown this to sometimes be the case when performing dynamic simulations of hybrid biofilm systems.
- The user wants to have direct control over the ammonia flux into the biofilm as well as the seeding effect as determined by the sloughing yield $Y_{Sloughed}$.

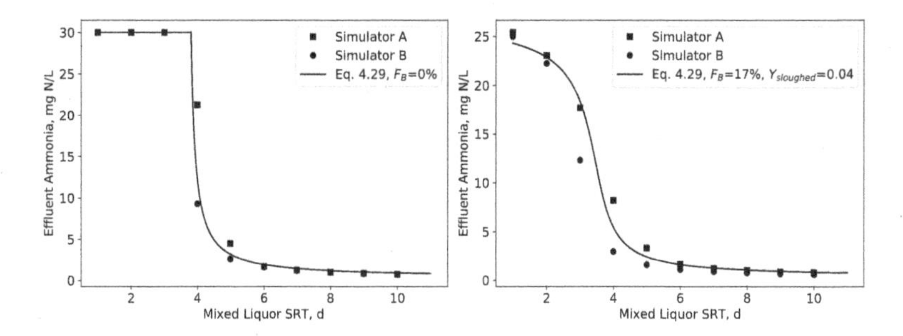

Figure 5.2 Comparison of washout curves using Simulator A: 1-d biofilm model, Simulator B: 2-compartment model, and Equation 4.29

> While it may be possible to calibrate a 1-d model to achieve the expected fluxes and seeding effect, in practice this can be extremely time-consuming and even impractical.

5.2.3 EQUIVALENCE OF 1-D AND 2-COMPARTMENT MODELS

The Case for 2-Compartment Models

Nitrification washout curves are presented in Figure 5.2 for a single, completely-mixed tank with and without MBBR media. The results are based on simulation performed using a 1-d model, a 2-compartment model, and Equation 4.29. Figure 5.2 demonstrates the basic equivalence between each of these modeling approaches.

In a third simulation package, Simulator C, a very different washout curve was initially simulated using default model parameters. This is presented in the "Default parameters" series in Figure 5.3 which indicates performance of the hybrid process was worse above the washout SRT than in the CAS process. The reason for this turns out to be that Simulator C predicts mixed liquor attachment to the biofilm, with higher attachment rates occurring at higher mixed liquor concentrations. As a consequence, thicker biofilms are predicted as a function of mixed liquor concentrations. Normally, 500 μm would already be considered too thick of a biofilm. Values of 1,000 μm and greater, which occur in Figure 5.3 at SRTs greater than 5 days, would simply not be sustainable in practice.

With a thick biofilm, Simulator C predicts a negative $Y_{Sloughed}$, *i.e.* negative seeding effect. This emergent behavior of the biofilm model is not realistic and is an artefact of the way the attachment/detachment model is set up. As shown in Figure 5.3, Simulator C can be calibrated to the behavior predicted by Equation 4.29 by lowering the default attachment rate. This also has the effect of maintaining the biofilm thickness in a more reasonable range. An important advantage of 2-compartment models is that it does not require this calibration step, which can be onerous when running simulations for multiple operational scenarios.

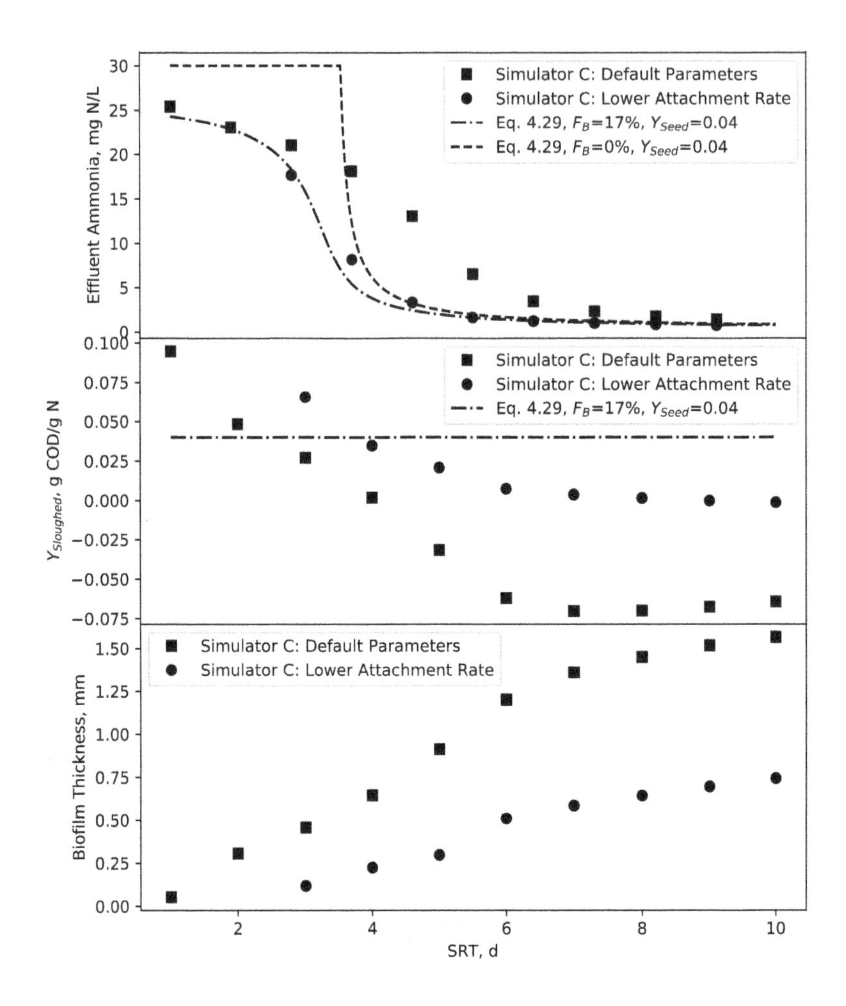

Figure 5.3 Variability in seeding effect as a function of biofilm thickness

The Case for 1-D Models

The above sections argue that 2-compartment models can provide equivalent predictions to 1-d models, that they are faster to simulate, simpler and more transparent in their use. However, in many cases, using a 1-d model is still preferable. For example, when a majority of the ammonia removal occurs in the biofilm, the 2-compartment model presented in Figure 5.1 might require very high rates of flow between the bulk liquid and biofilm. Simulating diffusion into the biofilm using bulk liquid flow may be an acceptable modeling simplification when the resultant flows are low, but likely not when they are high.

For dynamic simulations, where biofilm nitrification rates are expected to vary as a function of ammonia loading, BOD loading or some other factor, a 1-d model may also be more appropriate than a 2-compartment model. This will be explored in greater detail in the next chapter.

5.3 BIOKINETIC MODELS

ASM Models for Flocs and Biofilms

The so-called "ASM" biokinetic models were originally developed with the purpose of predicting sludge production, oxygen utilization, and nitrification in single-sludge, activated sludge processes [12]. They have been extended considerably over the years and even adopted to simulate the biology in biofilms. But not all of the modeling assumptions that are appropriate for activated sludge flocs are as appropriate for biofilms. Biofilms have unique properties. For example, biofilms are more likely to include more abundant, and different, higher life forms such as protozoa and nematodes. Abundance of EPS may not be the same either. ASM models do not include state variables for these and, as a result, their activity is implicitly accounted for, or "lumped into", the endogenous decay rate and endogenous yield of heterotrophic organisms. In light of this, one might expect separate parameter sets for endogenous decay to be applied to mixed liquor flocs and biofilms. Rightly or wrongly, the current best practice simulation of hybrid systems is to use identical parameter sets for both biofilm and mixed liquor organisms.

The application of ASM models to biofilms may also be questioned with respect to the use of Monod-style, half-saturation concentrations. For nitrification, the purpose of these half-saturation concentrations is to account for conditions where ammonia and oxygen become rate limiting. The half-saturation coefficients in ASM models implicitly account for some degree of diffusional mass transfer limitation since there is no transport model that can account for it explicitly. (Unlike biofilms, it is considered unnecessary to try to model 1-d structure and mass transport limitations in flocs.) For example, Table 5.1 shows that three of the leading process simulation software packages use a half-saturation coefficient for nitrification of 0.7 mg N/L, whereas bench-scale studies on pure cultures of nitrifying organisms would indicate that ammonia does not become rate limiting until concentrations are below 0.1 mg N/L. The higher value used in ASM models is to account for diffusional resistance into mixed liquor flocs. Using the same half-saturation coefficient in 1-d biofilms models amounts to "double counting". The diffusional resistance is already explicitly accounted for in 1-d biofilm models.

Heterotrophs and Nitrifiers

The two main organism groups in ASM models are heterotrophic and nitrifying bacteria. Accounting for heterotrophs in biofilms is useful as it relates to their competition with nitrifiers for oxygen or space. Too much heterotrophic activity will inhibit biofilm nitrification. In the absence of concerns around heterotrophic organism competition, the main focus of biofilm modeling in this book is understanding nitrification dynamics. The current state-of-the-art in commercial simulation software is

2-step nitrification modeling. 1-step models may be available, but in most cases their use is not being promoted by the software developers.

1-Step and 2-Step Nitrification Models

Initial publications of the ASM models always lumped the activity of ammonia oxidizing bacteria (AOBs) and nitrite oxidizing bacteria (NOBs) into a 1-step nitrification model [12]. Instead of being treated individually, AOBs and NOBs were generically referred to as "nitrifiers" or "autotrophs". The move to 2-step nitrification models is fairly recent and has come as a response to the commercialization of shortcut nitrogen removal technologies. In shortcut nitrogen removal, the activity of NOBs is suppressed and, owing to the opportunities it provides for energy and carbon savings, has quickly become the default strategy for treating high-strength wastewaters.

It is important to understand that, while achieving NOB suppression is relatively easy in high-strength, high-temperature wastewaters, it is rather difficult in the lower strength, lower temperature conditions that are typical of mainstream activated sludge. 2-step nitrification models, however, simulate NOB suppression rather easily in both environments. For this reason, they should be used with caution and relative abundance of AOBs and NOBs verified after simulation runs to evaluate whether NOB suppression is being simulated. If yes, it is worthwhile to ask whether this is really what is, or what may be, expected to be occurring in practice.

The above considerations are particularly relevant as concerns simulation of nitrification in oxygen-limited biofilms. The stoichiometric oxygen requirement for nitrification that stops at nitrite, NO_2^-, is 25% lower than when nitrification proceeds all the way to nitrate, NO_3^-. As illustrated in the reaction equations below, the model can either calculate oxygen utilization based on a stoichiometric ratio of 3.43 mg O_2/mg NH_4^+-N or (3.43 + 1.14) = 4.57 mg O_2/mg NH_4^+-N. For oxygen limited conditions, a model that simulates NOB suppression, that is nitrification that stops at nitrite, can result in 25% higher nitrification rates than a model that simulates complete nitrification to nitrate.

$$NH_4^+ + \frac{3}{2}O_2 \xrightarrow{\text{3.43 mg O/mg N}} NO_2^- + H_2O + 2H^+$$

$$NO_2^- + \frac{1}{2}O_2 \xrightarrow{\text{1.14 mg O/mg N}} NO_3^-$$

Table 5.1 presents default biokinetic parameters for nitrification from Simulators A, B and C. With some minor deviations, the major software providers use identical model structures and parameter values.

5.4 OXYGEN TRANSFER IN AN MABR

In conventional media supported biofilms, both ammonia and oxygen "co-diffuse" from the bulk liquid into the biofilm. In MABR biofilms, ammonia still diffuses from the bulk liquid, but oxygen is introduced inside the media lumen so that it can "counter-diffuse" from the inside of the biofilm towards the outside bulk liquid.

Table 5.1
Comparison of Selected Nitrification Biokinetic Parameters from Representative Simulation Software

Parameter	Units	Simulator A	Simulator B	Simulator C
μ	d^{-1}	0.9	0.9	0.9
b	d^{-1}	0.17	0.17	0.15
Y	d^{-1}	0.15	0.18	0.15
K_N	mg N/L	0.7	0.7	0.7
K_{DO}	mg O_2/L	0.25	0.25	0.25
θ_μ	–	1.072	1.072	1.072
θ_b	–	1.029	1.029	1.03

Modeling counter-diffusional biofilms is not very different from traditional, co-diffusional biofilms. A positive gradient of oxygen is simply introduced at the base of the biofilm. Calculation of the oxygen transfer across the MABR media into the biofilm can be approached in different ways. "Pressure-based" and "exhaust oxygen" based models are outlined below. Note that model jargon is to speak of "fluxes" into and out of the biofilm. What is otherwise referred to as oxygen transfer rate (OTR) in this book is referred to as an oxygen flux, J_{O_2}, in the sections below.

5.4.1 PRESSURE-BASED MODEL

The pressure-based approach was described by Côté in his seminal work on mass transfer in "bubble-free" aeration membranes [4]. In Equation 5.1, the flux of oxygen into the biofilm, J_{O_2} , is calculated based on a pressure difference between the inlet and outlet of the MABR media, $\left(\frac{P_{in}}{H} - \frac{P_{out}}{H}\right)$, relative to the driving force of oxygen into the biofilm. This driving force is a function of the oxygen concentration at the base of the biofilm, C_L, and the log average of the inlet and outlet oxygen partial pressures.

$$J_{O_2} = K \frac{\left(\frac{P_{in}}{H} - \frac{P_{out}}{H}\right)}{ln\left(\frac{\frac{P_{in}}{H} - C_L}{\frac{P_{out}}{H} - C_L}\right)} \tag{5.1}$$

Where:

J_{O_2}	:	Oxygen flux, $[ML^{-2}T^{-1}]$
K	:	Mass transfer coefficient, $[LT^{-1}]$
P_{in}	:	Inlet partial pressure of oxygen, $[ML^{-1}T^{-2}]$
P_{out}	:	Outlet partial pressure of oxygen, $[ML^{-1}T^{-2}]$
H	:	Henry's Law constant, $[L^2T^{-2}]$
C_L	:	Oxygen concentration in liquid, $[ML^{-3}]$

In Equation 5.1, the parameter with the greatest uncertainty and influence on fluxes is the concentration of oxygen at the base of the biofilm, C_L. The mass transfer coefficient, K, should be a known physical property of the MABR media, and the partial pressure of oxygen at the inlet and outlet to the media are easily measured. But the concentration of oxygen at the base of the media cannot be measured for most practical applications. It is a function of the "pull" of oxygen from biological respiration in the biofilm, and the "push" of oxygen, J_{O_2}, from the oxygen pressure inside the media lumen. Because J_{O_2} is itself a function of C_L, solving Equation 5.1 requires an iterative solution. This is easily handled using simulation software.

5.4.2 EXHAUST OXYGEN-BASED MODEL

An alternate approach to calculating oxygen flux, J_{O_2}, uses airflow and the difference between inlet and outlet oxygen purity as presented in Equation 5.2.

$$J_{O_2} = \frac{\left(Q_{air,in}x_{O_{2,in}} - Q_{air,out}x_{O_{2,out}}\right)\rho_{O_2}}{A} \tag{5.2}$$

Where:

J_{O_2}	:	Oxygen flux, $[ML^{-2}T^{-1}]$
$Q_{air,in}$	:	Media inlet airflow under standard conditions, $[L^3T^{-1}]$
$Q_{air,out}$	:	Media outlet airflow under standard conditions, $[L^3T^{-1}]$
$x_{O_{2,in}}$	:	Mole fraction of oxygen in the air inlet, $[-]$
$x_{O_{2,out}}$	:	Mole fraction of oxygen in the air outlet, $[-]$
ρ_{O_2}	:	Density of oxygen under standard conditions, $[ML^{-3}]$
F	:	Factor to account for volume loss between inlet and outlet, $[-]$
A	:	Surface area of media, $[L^2]$

In Equation 5.2, calculating oxygen flux, J_{O_2}, is dependent on two variables: the mole fraction of oxygen in the outlet, $x_{O_{2,out}}$, and the volumetric loss in airflow between inlet and outlet, $(Q_{air,in} - Q_{air,out})$. The mole fraction of oxygen in the inlet is assumed to be a constant 20.9% so long as atmospheric air is the source of air.

Volumetric loss of airflow in MABR's is a natural consequence of diffusion of nitrogen and oxygen gases across the media into the biofilm: the gases that diffuse across the media cannot contribute to airflow in the outlet. Diffusion of nitrogen will occur when the partial pressure of nitrogen inside the media lumen, and the concentration of dissolved nitrogen gas on the liquid side of the media, result in a positive gradient. Diffusion of oxygen will occur for the same reason, but is greatly enhanced by the respiration of nitrifying and heterotrophic organisms inside the biofilm. With higher respiration rates, there is more "pull" of oxygen from the media lumen into the biofilm. The consequence for Equation 5.2 is a widening gap in both $(x_{O_{2,in}} - x_{O_{2,out}})$ and $(Q_{air,in} - Q_{air,out})$. Accounting for differences in the former, but not the latter, will result in underestimation of flux, J_{O_2}.

Volumetric loss can be accounted for using a factor, F_V, based on a simple oxygen balance according to Equation 5.3.

$$F_V = \frac{1 - x_{O_{2,in}}}{1 - x_{O_{2,out}}} \tag{5.3}$$

For a conventional air application where inlet oxygen, $x_{O_{2,in}}$, is 20.9%, the oxygen transfer efficiency (OTE) can be calculated solely based on measured outlet, or exhaust, oxygen purity as follows:

$$OTE = \frac{20.9\% - F_V x_{O_{2,out}}}{20.9\%} \tag{5.4}$$

Equation 5.4 provides a useful basis for calculating OTE using a single measured value: outlet oxygen concentration. It is important to note that, because Equation 5.3 ignores the nitrogen diffusion into the biofilm, the result is underestimation of the volumetric loss factor, F_V, as well as oxygen transfer in Equation 5.4. None of the commercial process modeling packages that include MABR models account for nitrogen diffusion either. This is important to remember when considering model calibration to field data.

5.5 CLOSING THOUGHTS ON SIMULATION SOFTWARE

Whereas design of the activated sludge process using commercial simulation software is widespread in the industry, engineers have been more hesitant to rely on these tools for biofilm processes. In the words of one skeptical process engineer: "unlike activated sludge models, the biofilm models just aren't there". This attitude may reflect thinking that, since biofilms are complex, modeling them must be complex, too. But this ignores that, in many cases, simple models are able to describe the mechanisms governing the behavior of interest. There might be countless mechanisms that contribute to biofilm structure and biological activity, but if only one of these is required to explain nitrification, then the system is not that complex at all, at least not with respect to the behavior of interest. It has long been known that diffusion is the mechanism governing nitrification rates in nitrifying biofilms, and diffusion models can be quite simple indeed.

But how well do 1-d models predict whether or not a biofilm is a nitrifying biofilm? In fact, where 1-d biofilm models seem to be most challenged is in accounting for factors that can impact viability of nitrifiers in the biofilm: heterotrophic organism competition, biofilm thicknesses and sloughing / detachment rates. Where nitrification is impacted by these processes, model predictions are likely to be treated with less confidence. Nevertheless, model-based sensitivity analyses may provide useful insights into which conditions are more likely to lead to nitrifying biofilms, NOB suppression or a prohibitively thick biofilm.

Chapter 4 presented design equations to quantify the performance of hybrid processes as a function of ammonia removal in the biofilm, $F_{Nit,B}$, and the seeding effect, $Y_{Sloughed}$. The predictive power of these equations, however, was limited to the assumption of a complete-mix activated sludge bioreactor and steady-state conditions. Simulation software allows these limitations to be overcome: performance of hybrid processes can be evaluated for any process configuration, even biological nutrient removal (BNR) configurations which involve multiple aerated and unaerated zones. Furthermore, as will be explored in the next chapter, simulation software can evaluate dynamic conditions.

Although models that deliver accurate predictions of nitrification rates in biofilms are desirable, this is not "mission critical" to understanding the role of biofilms in hybrid processes. In fact, the 2-compartment model presented in Section 5.1.4 does not predict nitrification rate at all; it is set by the user. What is important is that, for any given biofilm nitrification rate, its effect can be evaluated within the framework of the overall activated sludge model. This enables useful predictions of hybrid system performance which will be explored further in the next chapter.

6 Investigation of Operational Dynamics

Steady-state design equations were developed in Chapter 4 to quantify how removing some fraction of the influent load in the biofilm, as well as the seeding effect, leads to activated sludge process intensification. This chapter takes the argument for the process intensification benefits of hybrid systems one step further by exploring dynamically varying, *i.e.* time-dependent, conditions. There can be many sources of this variability but the focus here will be on influent loading dynamics.

One of the overlooked benefits of hybrid systems is that natural load dampening, or peak shaving, is provided by biofilms operated under ammonia-limited conditions. The reason for this is that biofilm activity is diffusion, or gradient, limited. The biofilm itself might contain more than enough nitrifying organisms to treat the incoming load, or at least it might if they were nitrifying at their maximum potential rate. But, as will be discussed, biofilms tend to operate below their potential under average loading conditions. This makes it possible for them to "rise to the occasion" when the peak loading events occur.

The key to achieving these load dampening benefits really is ammonia-limited conditions. In conventional biofilms, oxygen-limiting conditions tend to exist, so no increase in nitrification rate occurs under higher ammonia loading. In MABR biofilms, however, non-limiting oxygen conditions can readily be achieved, and so ammonia diffusion is rate limiting to nitrification. As a result, these biofilms increase their nitrification rate as the bulk ammonia concentration, *i.e.* loading, increases. The increase in biofilm nitrification rates under peak loading conditions naturally attenuates, or dampens, the load that needs to be nitrified in the mixed liquor. This is an important benefit of MABR/AS processes that was not accounted for in the steady-state design curves of Chapter 4. The load dampening benefits of MABR will be demonstrated using field data from the recently commissioned MABR/AS process at the Yorkville-Bristol Sanitary District (YBSD) in Illinois, U.S.A. A case study will be used to assess dynamic process performance for the following scenarios:

- Conventional Activated Sludge (CAS),
- IFAS with oxygen-limited biofilm, and
- MABR/AS IFAS with ammonia-limited biofilm.

The approach to model calibration and performance evaluation presented in this chapter can be extended to other IFAS configurations whether the support media be aerated, MABR, or not, MBBR and Fixed-media.

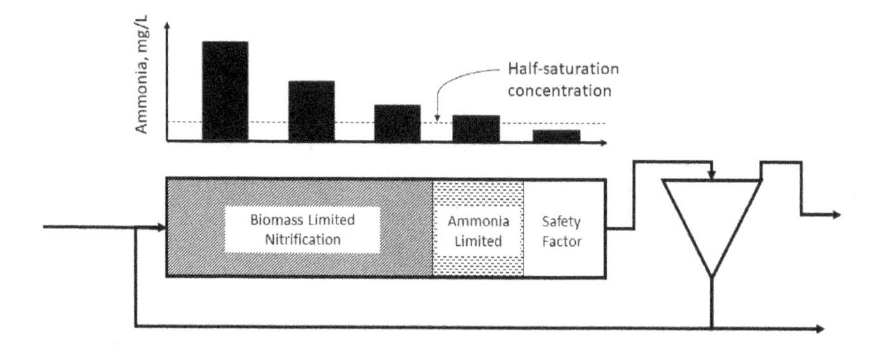

Figure 6.1 Schematic representation of ammonia profile in a conventional activated sludge (CAS) bioreactor with impacts on nitrification rates

6.1 MANAGING LOADING DYNAMICS

6.1.1 CONVENTIONAL ACTIVATED SLUDGE

As discussed in Chapter 1, the ability to manage loading variation is accounted for in conventional process design by applying a safety factor to the minimum SRT. This increases the inventory of nitrifying organisms over and above what is required to treat the average load. As a result, under average loading conditions, there is insufficient ammonia to sustain nitrification at the maximum potential rate. The mass of nitrifiers remove all, or almost all, of the average influent load, but their overall rate of activity is not as high as it could be. The inventory of nitrifying organisms is said to be ammonia-limited.

As illustrated in Figure 6.1, this will manifest itself in a plug-flow bioreactor in relation to the profile of ammonia between the upstream and downstream ends of the bioreactor. Upstream in the bioreactor, ammonia concentrations are well above the half-saturation concentration, so non-limiting, and the mass of nitrifiers in this zone is removing ammonia at its maximum rate. Towards the downstream end of the bioreactor, ammonia concentrations approach the half-saturation concentration, and activity of the nitrifiers becomes ammonia-limited. When nitrification is ammonia-limited, it means that nitrification rates will increase when loading increases, and this helps to minimize the breakthrough of ammonia into the downstream zone of the bioreactor. This downstream zone, where ammonia concentrations are below the effluent objective, is labeled in Figure 6.1 as "Safety Factor". During peak loading events, it acts as a buffer zone to prevent ammonia breakthrough from reaching the effluent.

Even though, in Figure 6.1, only the downstream portions of the bioreactor are "ammonia-limited", because it is the same sludge that is recycled through the different zones, "ammonia-limited" conditions apply to the activated sludge as a whole.

Ammonia-Limited Activated Sludge

For suspended growth mixed liquor, an ammonia-limited condition is desirable because it means that, when peak loading events occur, nitrification will increase to match the incoming load. In other words, if more ammonia were available, more ammonia would have been removed. This only holds true, however, up until the point at which the maximum potential nitrification activity of the sludge inventory has been reached. Beyond this point, any increase in influent ammonia loading will pass directly through to the effluent. So, by multiplying the inventory of nitrifying organisms in the bioreactor, applying a safety factor to SRT can increase the maximum potential nitrification activity, and lessen the risk of ammonia breakthrough into the effluent. It follows from this that ammonia-limited conditions are fundamental to achieving safety factor in activated sludge. This is an important point to remember when considering hybrid biofilm processes and the relative benefits of ammonia-limited *vs.* oxygen-limited biofilms.

As in Chapters 1 and 4, the maximum potential nitrification activity will be discussed in this chapter in terms of the parameter $R_{Nit,max/L}$. Once nitrification activity has reached $R_{Nit,max/L}$, nitrification is "biomass limited", and there is no more ability to treat any increase in the influent load.

Besides ammonia-limited and biomass-limited conditions, there is also the possibility that nitrification can be "oxygen-limited". This condition is avoidable in suspended growth mixed liquor by maintaining sufficient bulk dissolved oxygen concentration, typically 2 mg/L or higher. Diurnal "DO sags" are not uncommon, however, when the aeration systems are unable to meet the oxygen demand of the mixed liquor. DO sag can easily lead to ammonia breakthrough in the effluent.

A Useful Proxy for Safety Factor

Back in Chapter 1, the influence of SRT on effluent ammonia concentration S_{NHx}, mass of nitrifiers in the bioreactor M_{AOB}, and the ratio of maximum potential nitrification activity to influent loading $R_{Nit,max/L}$ were discussed. The relationships were summarized as a series of curves in Figure 1.2 illustrating the principle that a higher $R_{Nit,max/L}$ can be achieved by operating at longer SRTs. This parameter provides a nice means for quantifying the benefit of safety factor: safety factor enables treatment of higher "peak" ammonia loads according to the ratio $R_{Nit,max/L}$.

It is important to note, however, that the increase in $R_{Nit,max/L}$ is not linear with respect to SRT. This is due to the role of endogenous decay, which has a greater impact on nitrifier inventory at longer SRTs. An engineer might select a design SRT that is twice the minimum SRT and declare a safety factor of two. However, the ability to handle incoming peak loads, as quantified by $R_{Nit,max/L}$, would not be doubled, and the reason for this is because of the increased endogenous decay at the longer SRT. The parameter $R_{Nit,max/L}$ can be thought of as a proxy for safety factor that provides a more accurate assessment of the ability to handle loading peaks. An additional benefit is that, as discussed in Chapter 4, $R_{Nit,max/L}$ can be calculated for hybrid systems. It therefore provides a useful basis for comparing the safety factor in CAS and IFAS processes.

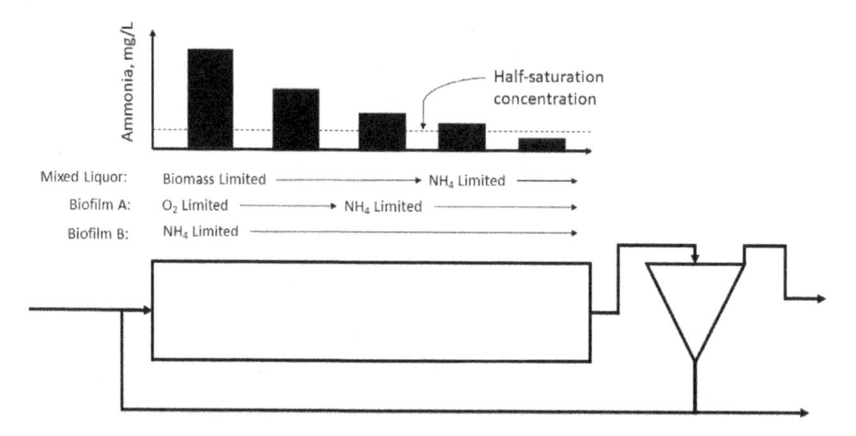

Figure 6.2 Schematic representation of ammonia profile in a hybrid bioreactor where Biofilms A and B represent co-diffusional (Fixed-media, MBBR) and counter-diffusional (MABR) biofilms, respectively

6.1.2 IFAS SYSTEMS

Just like in conventional activated sludge, the response of biofilms to dynamic loading conditions can be assessed with respect to whether they are ammonia-limited, oxygen-limited or biomass-limited. These conditions will influence process behavior as follows:

- Nitrification rate increases with increased loading. This will occur in biofilms that are ammonia-limited.
- Nitrification rate remains unchanged with increased loading. This can occur in biofilms that are insensitive to changes in bulk ammonia concentration because they are either oxygen- or biomass-limited. In either case, the result can be a biofilm that is "fully penetrated" with ammonia and thus insensitive to changes in bulk liquid concentration.
- Nitrification rate decreases under increased loading. This might occur in oxygen-limited biofilms where nitrification is inhibited by heterotrophic activity and therefore is sensitive to variation in BOD loading, which may be increasing at the same time as ammonia loading.

Ammonia-Limited Hybrid Systems

The bioreactor ammonia profile, and its impacts on nitrification kinetics, is shown for a hybrid system in Figure 6.2 showing the transitions from biomass-limited, or oxygen-limited, to ammonia-limited conditions in both the mixed liquor and biofilm. Two different types of biofilms are included, Biofilm A and B, to represent two different scenarios with respect to oxygen availability. For Biofilm A, oxygen diffusion into the biofilm is limiting to nitrification as long as the ammonia concentration is above some threshold. Where this threshold occurs, however, may vary according to bulk dissolved oxygen concentrations, for conventional biofilms, or lumen air pressure, for MABR biofilms. In contrast, for Biofilm B, oxygen is never limiting to nitrification and so ammonia-limited conditions are achieved for the entire bioreactor. Note that, because biofilms are almost always diffusion-limited processes, biomass-limited conditions are rarely encountered and so this case is not presented for either Biofilm A or Biofilm B.

As discussed in Section 6.1.1, ammonia-limited conditions provide the ability to manage peak loading events. As a result, a hybrid process consisting of a Biofilm + Mixed Liquor would be expected to better manage peak loading events than Mixed Liquor alone. And a hybrid process including Biofilm B would be expected to be better than Biofilm A. These hypotheses will be further explored through modeling in Section 6.3.

Avoiding Oxygen Limited Biofilms

If ammonia-limited biofilms really are preferable for managing peak loads, then the natural question is how to achieve this? In short, by avoiding oxygen-limited conditions. For MBBR and conventional Fixed-media processes, this is addressed by operating at as high a bulk liquid dissolved oxygen concentration as possible. Typical design guidelines recommend a dissolved oxygen range of 3 to 4 mg O_2/L [9]. Above this range, energy requirements for aeration become prohibitive. Another strategy is to avoid placing biofilm support media in the upstream zones of the bioreactor where BOD loading is highest. In these upstream zones, heterotrophic organisms are more likely to outcompete nitrifying organisms for dissolved oxygen inside the biofilm.

Avoiding placement of biofilm support media in the upstream zones of the bioreactor, however, results in a missed opportunity to maximize the zone of ammonia-limited nitrification. This is where MABR media provides a distinct advantage. As discussed in Chapter 2, Section 2.2.3, high oxygen diffusion rates into MABR biofilms can be achieved by tuning the lumen air pressure. In addition, the counter-diffusional nature of the biofilm gives nitrifying organisms an advantage in outcompeting heterotrophic organisms for the available oxygen. Whether MABR biofilms achieve ammonia-limited behavior in practice will be explored in the next section.

6.2 MABR BIOFILM DYNAMIC BEHAVIOR

6.2.1 FIELD DATA FROM YBSD

The MABR/AS case study for the Yorkville-Bristol Sanitary District (YBSD) was presented in Chapter 3, Section 3.3.1. The layout of this site as well as the location of the MABR media in Tank 2 of the 10-tank layout was presented in Figure 3.2. In this chapter, plant performance data will be explored in greater detail to demonstrate the ammonia-limited behavior, and peak load trimming benefits, of the MABR biofilm at this site.

Figure 6.3 presents flow, ammonia concentrations, ammonia removal and exhaust oxygen purity for the MABR zone. Flow through the MABR zone is calculated by

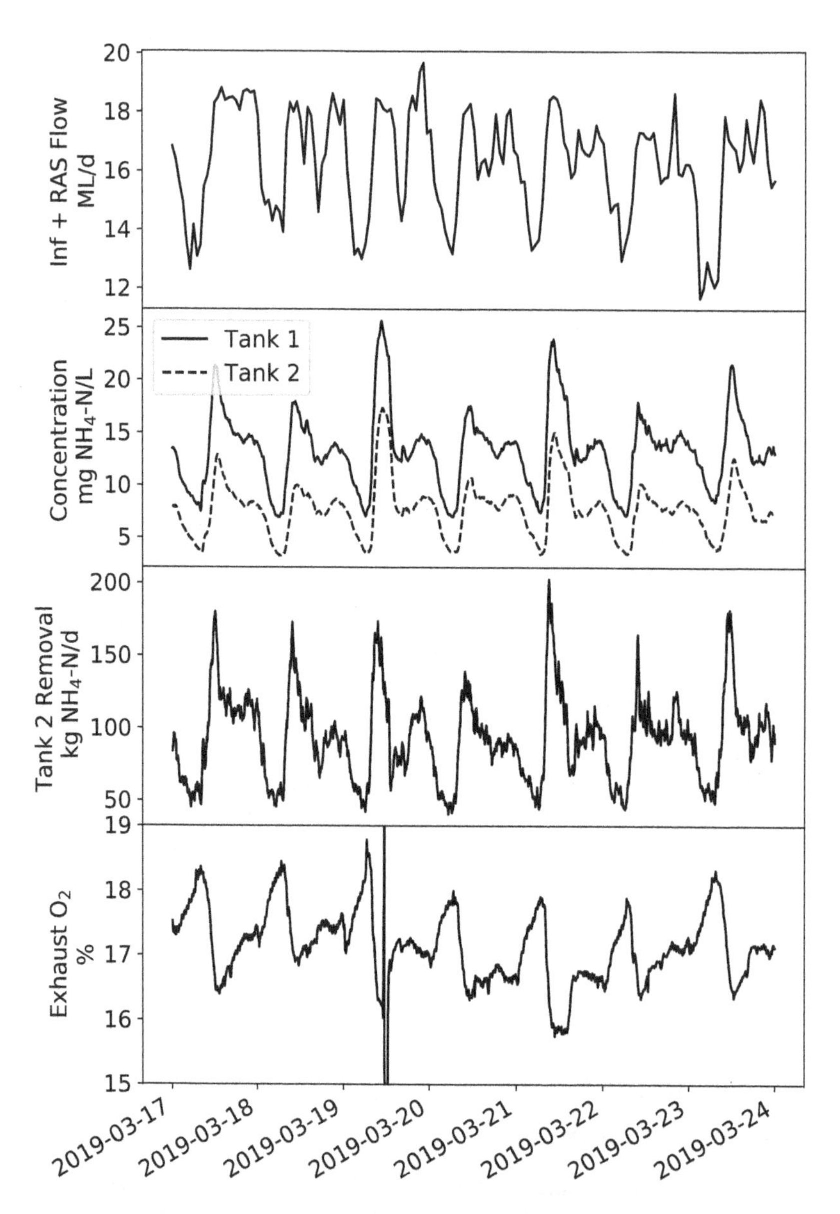

Figure 6.3 Field data of MABR performance from YBSD

adding the flow meter readings for the screened sewage and return activated sludge (RAS). Exhaust oxygen is based on the readings from an online gas meter that receives the collected gas from all 12 ZeeLung cassettes. The ammonia concentrations in Tanks 1 and 2 are based on readings from ion selective electrode (ISE) ammonia sensors immersed in the tanks. The ammonia removal in the MABR zone is calculated based on the difference between the concentrations in Tanks 1 and 2, and the combined screened sewage and RAS flows.

The reliability of the measurements presented in Figure 6.3 is not the same. Flow metering at this site is considered reliable as are the MABR exhaust oxygen measurements. Comparison of the ammonia ISE sensor data with laboratory analyses of grab samples, however, has shown significant deviation attributable to challenges in achieving, and maintaining, good sensor calibration. Quantification of ammonia removal across the MABR zone at this site has therefore been based on laboratory analysis of grab or composite samples from Tanks 1 and 2.

Given the doubts around the accuracy of the ammonia sensor data presented in Figure 6.3, the value this data provides is in characterizing trends, rather than quantifying absolute values. The difference between the ammonia concentrations measured in Tanks 1 and 2, $\Delta_{(1-2)}$, may not be accurate in absolute terms, but the diurnal variation in $\Delta_{(1-2)}$ still provides valuable insights into the ammonia-limited behavior of the MABR biofilm. The data from Figure 6.3 indicates a repeatable trend of increase in ammonia removal, both on a concentration and load basis, during the periods of the day when ammonia concentration in the MABR zone is highest.

Ammonia Removal vs. Bulk Ammonia Concentration

The time-series data from Figure 6.3 is presented in Figure 6.4 as scatter plots of ammonia removal *vs.* bulk ammonia concentration in the MABR zone, Tank 2, for four consecutive days in March 2019. These scatter plots provide a means for validating the hypothesis of ammonia-limited biofilm behavior: increasing bulk ammonia concentrations should correlate with increasing ammonia removal, which they do. On each of the four days, the correlation is between ammonia and concentration is clear but shows signs of hysteresis: the relationship follows a different path on the way up than it does on the way down.

Two factors should be considered when assessing the relationships presented in Figure 6.4:

- The hydraulic residence time in the MABR zone is approximately 45 minutes and so removals will be positively biased when Tank 1 ammonia concentrations are increasing, and negatively biased when they are decreasing.
- The y-axis variable is not independent of the x-axis variable: Tank 2 ammonia concentrations are used to calculate the ammonia removal across that zone.

The first point likely explains the hysteresis observed in Figure 6.4 whereas the second may undermine, to some degree, the confidence in the observed correlations.

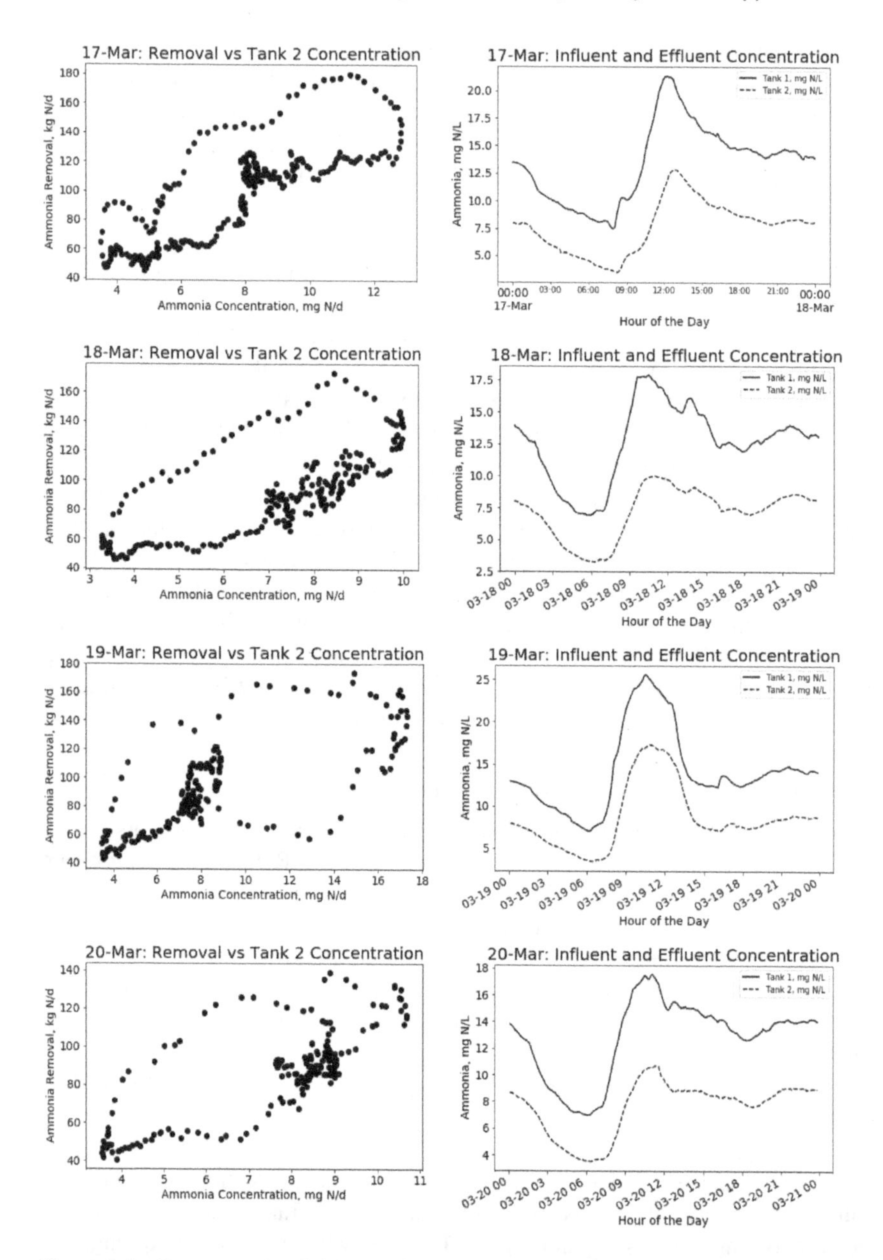

Figure 6.4 Representative field data of ammonia removal *vs.* bulk liquid ammonia plots from YBSD for Winter, Spring, Summer and Fall

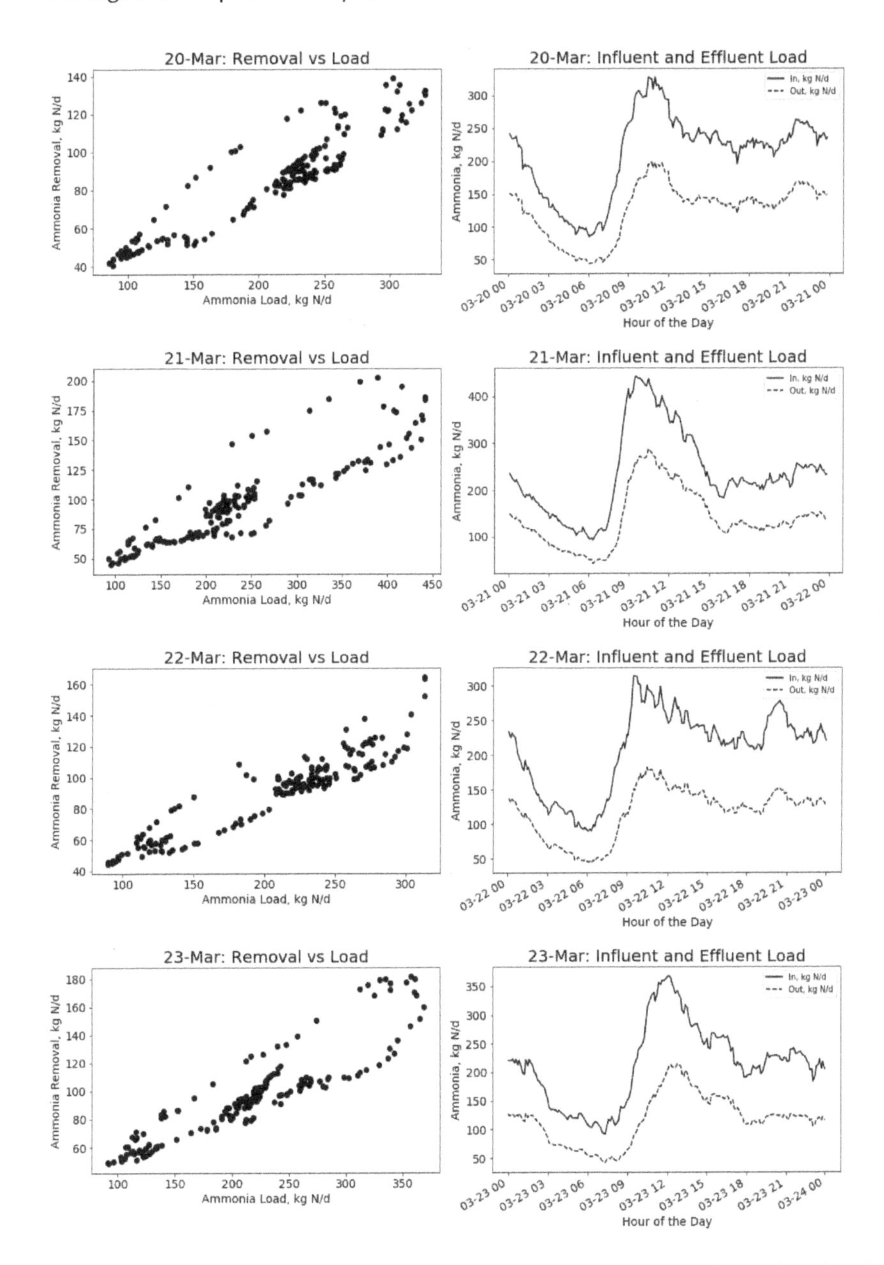

Figure 6.5 Representative field data of ammonia removal *vs.* ammonia loading plots from YBSD for Winter, Spring, Summer and Fall

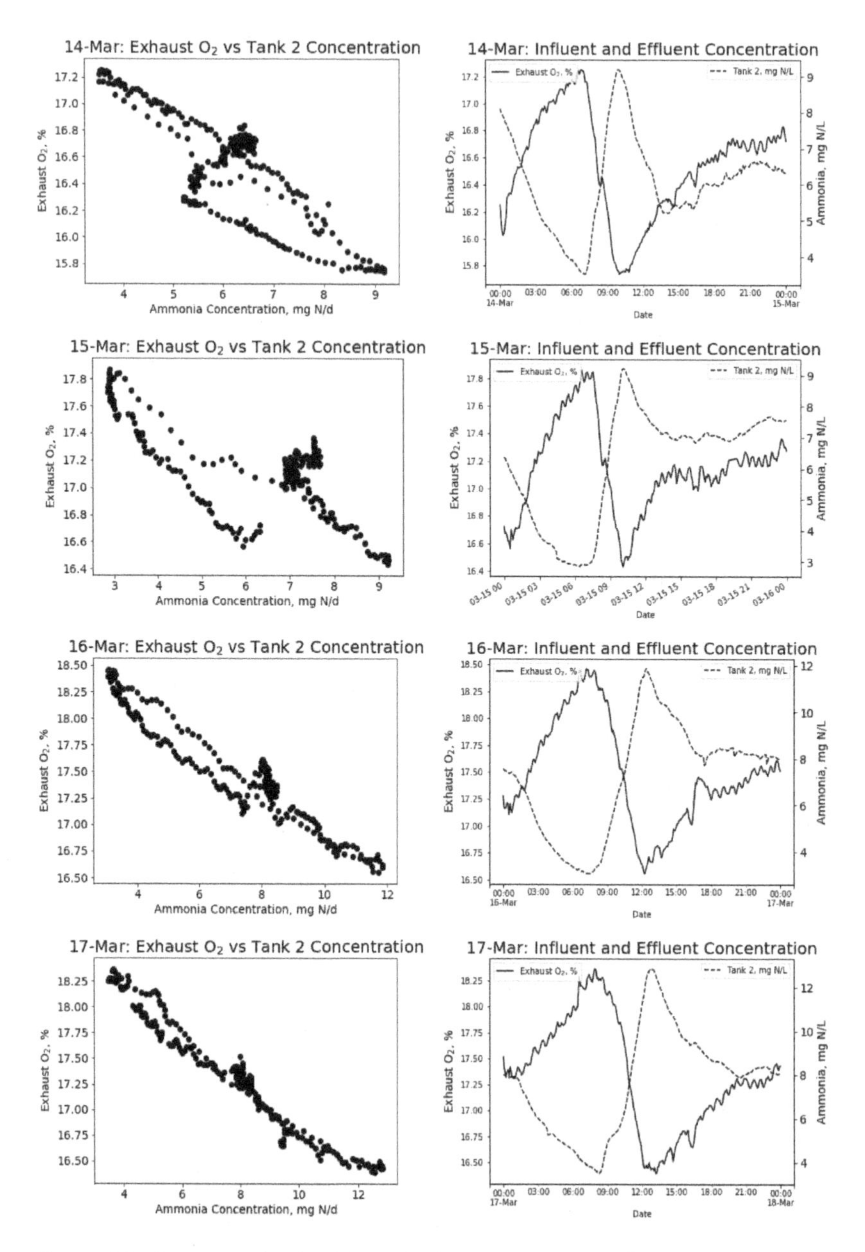

Figure 6.6 Representative field data of exhaust oxygen *vs.* bulk liquid ammonia plots from YBSD for Winter, Spring, Summer and Fall

Ammonia Removal vs. Loading

An alternate way to treat the data from Figure 6.3 is to plot ammonia removal *vs.* ammonia loading. This is presented in Figure 6.5. What is interesting about this figure is that the correlation between removal and loading becomes even stronger than when plotting removal *vs.* concentration. In addition, the signs of hysteresis observed in Figure 6.4 are not as strong. This is counter-intuitive because the *nitrification rate in a biofilm does not respond directly to loading but rather to the bulk ammonia concentration in the surrounding mixed liquor.* There could be circumstances, for example during wet weather events, when influent peak loads might not correlate with high bulk ammonia concentrations and diffusion gradients into the biofilm. Why then the stronger correlation to loading?

The explanation for the improved correlation between removal and loading, as compared to removal and bulk concentration, is not entirely clear. It can be noted, however, that the y-axis and x-axis variables are even less independent in removal *vs.* loading than they are for removal *vs.* concentration. Whereas in Figure 6.4 the y- and x-axes are both functions of a single variable, the Tank 2 ammonia concentration, in Figure 6.5 the y- and x-axes are both functions of two variables, flow and Tank 1 ammonia concentration. The greater interdependance between the y- and x-axes in Figure 6.5 may explain the stronger correlation.

Exhaust Oxygen vs. Bulk Ammonia Concentration

Plotting exhaust oxygen *vs.* bulk ammonia concentration provides a third means to validate the hypothesis of ammonia-limited biofilm behavior: increasing bulk ammonia concentrations should correlate with decreasing ammonia removal, which they do. As presented in Figure 6.6, the correlation between exhaust oxygen and bulk ammonia concentration is very strong. The signs of hysteresis are strong on some days, but not others.

The scatter plots from Figure 6.6 provide a particularly attractive means for quantifying the response in biofilm nitrification to peak loading events for the following reasons:

- Both the x- and y-axes variables are completely independent so there is no reason to believe that correlation is due to variable interdependence.
- The plots are based on a more reliable sensor, oxygen purity, that is less susceptible to drift or calibration errors than ISE ammonia sensors.
- Only one ammonia sensor is required, which reduces the cost and maintenance burden as compared to relying on two sensors to calculate ammonia removals.
- Exhaust oxygen measurements can be used to calculate oxygen transfer efficiency and rates, and even nitrification rates.

The calculation of nitrification rates from exhaust oxygen data requires characterization of the ratio of oxygen transfer to nitrification rate. For conventional nitrification, the stoichiometric ratio is 4.57 kg O_2/kg NH_4-N. For YBSD, the ratio is

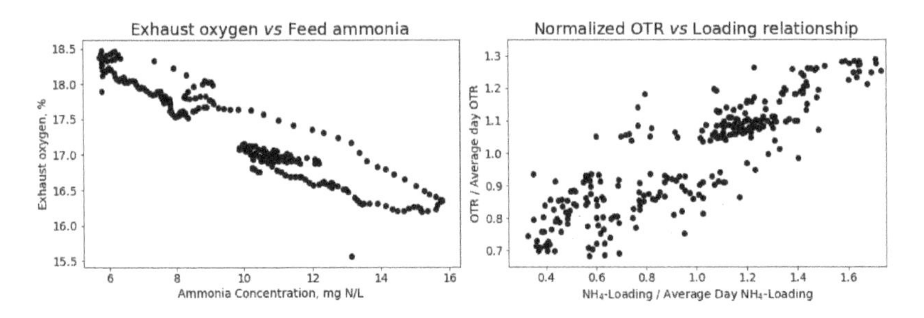

Figure 6.7 Diurnal response of oxygen transfer rate to ammonia loading based on field data from YBSD on November 9th, 2018

calculated using the measured nitrification rate, based on composite samples from Tanks 1 and 2, and oxygen transfer rates, based on the MABR airflow and exhaust oxygen purity. The resulting ratio tends to be in a range of 5 to 7 kg O_2/kg NH_4-N, where a higher ratio is indicative of more of the oxygen in the biofilm being directed to heterotrophic, BOD oxidizing activity.

Peak Load Shaving

Figures 6.4, 6.5 and 6.6 validate the ammonia-limited condition of the MABR biofilm at YBSD. This ammonia-limited behavior provides load dampening, or peak shaving, at a location upstream in the bioreactor where it could not be achieved conventionally. This translates to greater safety factor. On their own, however, these figures do not quantify how much load dampening the MABR biofilm actually provides to the overall process.

To provide this quantification, data from 9 Nov 2018 are presented in Figure 6.7 to show normalized variation in oxygen transfer rate relative to variation in ammonia loading. Oxygen transfer rates, and therefore nitrification rates, are seen to vary by +/- 30% as ammonia loading varies by +/- 60%. This is beneficial behavior for the stability of the overall process because it effectively means that the biofilm does less when the loading is low, thereby "feeding" the downstream suspended growth nitrifying organisms. When the loading is high, however, the biofilm nitrifies at a higher rate. The resultant load dampening at the upstream end of the process helps to avoid overloading the suspended growth organisms at the downstream end. In comparison, as illustrated in Figure 6.1, mixed liquor at the upstream end of the process is never ammonia-limited and thus provides no load dampening at all.

There is potential to further increase the load dampening benefits of an MABR biofilm by modulating the airflow to create oxygen-limited conditions under low loading conditions. This could be based on timer control, assuming that low loading conditions occur during predictable hours of the night and early morning, or using feed-forward or feedback ammonia-based aeration control (ABAC). This is yet another way in which biofilms can help manage peak loading events, and thereby enhance safety factor in the activated sludge process. But unlike seeding effect, which can be quantified using the design equations presented in Chapter 4, Section 4.3.4, the load-dampening benefits of an ammonia-limited biofilm are best quantified using dynamic process modeling.

The benefits of throttling the airflow to the MABR to limit biofilm nitrification rates, even during low loading periods, may not seem intuitive. After all, one would presumably want to maximize performance of the biofilm at all times. However, in a hybrid system there can be benefit to allowing more ammonia to "feed" the mixed liquor when the plant is underloaded, particularly when peak load dampening, or shaving, is an operational priority. This ensures that the mass of nitrifiers in the mixed liquor inventory is maximized, while allowing the biofilm to work when it is needed most, under the peak loading condition.

6.2.2 SIMULATING FIELD DATA

Oxygen Transfer Model Selection

To accurately capture dynamic behavior of MABR in a model, careful attention must be paid to the simulated response of the biofilm to bulk ammonia concentrations. Without this, the load dampening effects of MABR biofilms will be ignored. Chapter 5 presented several approaches to modeling oxygen transfer into counter-diffusional "MABR" biofilms. The simulation results presented in the following sections were generated using a commercially available model which relies on the following inputs:

- mass transfer coefficient for oxygen through the media wall,
- boundary layer thickness for diffusion of soluble material between the bulk mixed liquor and the biofilm, and
- pressure difference between the inlet and the outlet of hollow fiber MABR media.

This model was selected because, in contrast to other approaches, it allows exhaust oxygen to be calculated dynamically rather than being set as a model input. This is important because exhaust oxygen is one of the most critical pieces of field data characterizing performance. Matching model predictions to field data is the basis of model calibration, and exhaust oxygen has proven to be the most reliable source of field data.

Oxygen Transfer Model Calibration

Just like with the seeding effect, discussed in Chapter 5, oxygen transfer model parameters need to be carefully adjusted to correctly simulate the key behavior. What constitutes "key behavior" may vary on a case-by-case basis, but the dynamic relationship between oxygen transfer and ammonia concentration is a good place to start. This relationship is demonstrated by the field data from YBSD presented in Figure 6.3, and then again in Figure 6.6. A similar relationship would be expected for most municipal applications where oxygen utilization in the biofilm is primarily directed to nitrification.

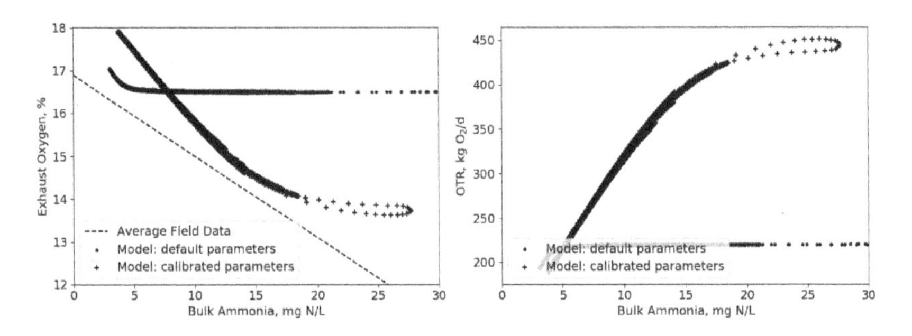

Figure 6.8 Simulated exhaust oxygen *vs.* bulk liquid ammonia plots assuming default and high mass transfer coefficient

Simulation results of exhaust $\%O_2$ *vs.* bulk ammonia concentration are presented in Figure 6.8 for oxygen-limiting and ammonia-limiting conditions. The software package's default mass transfer coefficient actually leads to oxygen-limiting conditions, characterized by no change in exhaust oxygen as a function of bulk ammonia concentration, *i.e.* a horizontal line. Calibration of the mass transfer coefficient enables the relationship from the field data to be captured to a reasonable degree.

6.3 MABR/AS CASE STUDY

Fully understanding, and indeed quantifying, the contribution of the biofilm to performance of a hybrid biofilm/activated sludge process may not always be obvious. Side-by-side operations of hybrid and conventional activated sludge process trains, under identical loading conditions, would be nice but is rarely feasible in practice. So process modeling is invariably the tool used by engineers to evaluate the benefits of hybrid processes, as compared to the conventional alternative.

The following sections provide an example of how this might be done for a CAS process that has been retrofit with MABR media to offset the loss of nitrification "safety factor" after conversion from a fully-aerobic process, to a BNR process with 40% unaerated mass fraction. The principles described below could be applied to any number of alternate process configurations.

6.3.1 MODEL SETUP

The case study presented here compares the response to loading dynamics of the (a) CAS and (b) MABR/AS process configurations presented in Figure 6.9. Both configurations consist of five bioreactor tanks in series followed by secondary clarification with return activated sludge. Tank 1 is designated as an anaerobic zone and Tank 2 is an anoxic zone. In the MABR/AS alternative, MABR biofilm support media is installed in Tank 2. The result is a hybrid zone in which the biofilm is aerobic but the surrounding mixed liquor is anoxic.

For the MABR/AS configuration, Tank 2 contains sufficient media surface area to nitrify approximately 30% of the influent ammonia load. Tanks 3, 4 and 5 are fine bubble aeration zones that provide a 60% aerobic volume fraction. This configuration was chosen because it is one of the potential flow configurations for the YBSD plant described in Section 3.3.1. Also, in the absence of MABR media in Tank 2, the configuration looks like a fairly typical A^2O process, appropriate for biological phosphorus removal.

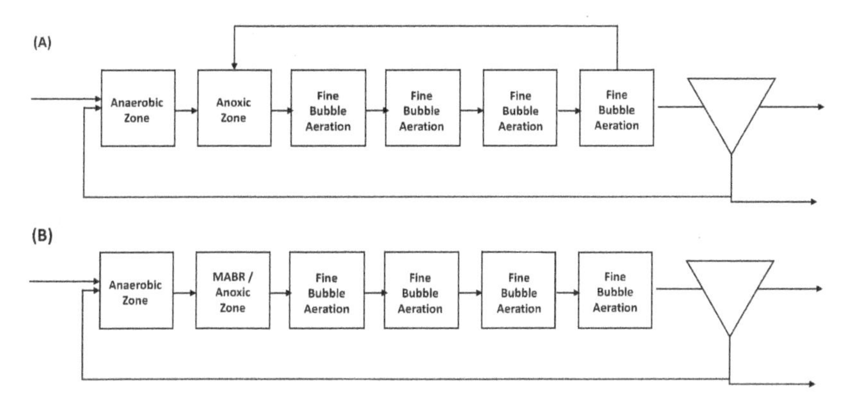

Figure 6.9 Process configuration of five tanks-in-series as (a) conventional A^2O process and (b) MABR/AS process

Loading Dynamics

The influent flow and loading patterns used in model simulations are presented in Figure 6.10. These patterns were developed based on 10-minute-interval data from YBSD. Flows are based on readings from influent flow meters. The influent ammonia load and concentrations were calculated based on data from an ion selective electrode (ISE) ammonia sensor located in Tank 1 and influent and RAS flows according to the following formula:

$$NH_{4,Inf} = NH_{4,Tank1} \frac{\left(Q_{Inf} + Q_{RAS}\right)}{Q_{Inf}} \tag{6.1}$$

The resulting pattern is characterized by maximum daily and hourly flow peaking factors of 1.25 and 1.57, respectively, and maximum daily and hourly ammonia load peaking factors of 1.46 and 3.64, respectively.

The diurnal variation in flow and load observed in Figure 6.10 are, for the most part, as would be expected based on normal domestic water use patterns. However, there are notable spikes in the hourly ammonia loading that occurs on days 29, 30 and 31. Such day-to-day variation is typically attributable to changes in industrial or commercial water use in the sewershed or plant internal recycle loads, for example sludge dewatering return flows.

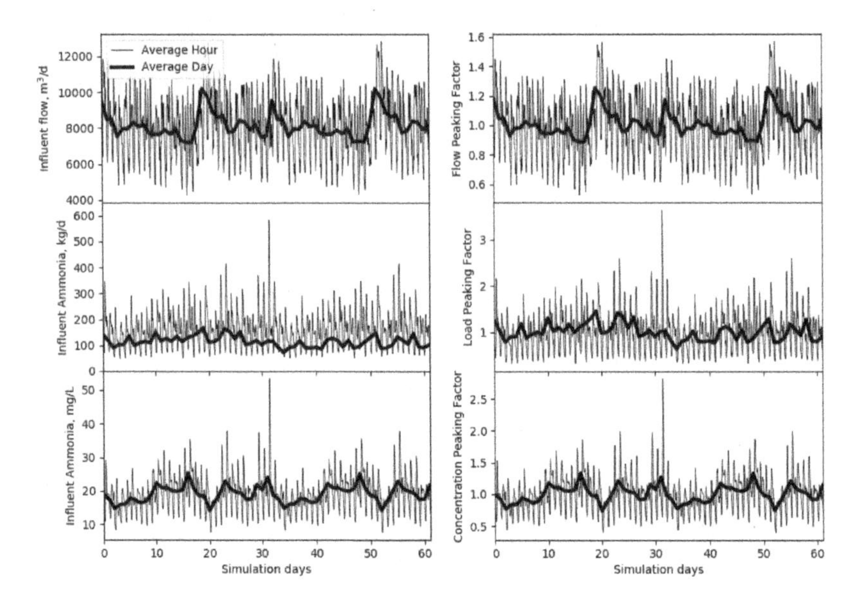

Figure 6.10 Influent flow and loading profile used in dynamic simulations

Basis of Model Calibration

The previous section described the importance of the relationship between exhaust oxygen and bulk ammonia concentration to load dampening, *i.e* ammonia-limited behavior. Dynamic process modeling must be able to capture this ammonia-limited behavior. In addition, as discussed in Chapter 5, it is important to pay attention to biofilm thickness. Too high or too low of a biofilm thickness can lead to unreasonable predictions of "seeding effect". A biofilm thickness of approximately 200 to 300 μm is a good range to represent what can be observed in field conditions, and results in a seeding effect characterized by a sloughing yield, $Y_{Sloughed}$, or 0.05 mg COD/mg NH_4-N. Ammonia-limited behavior and controlling biofilm thickness are the sole basis of model calibration for the results presented in the following sections.

Summary of Model Inputs

Design criteria for the process configuration presented in Figure 6.9 are summarized in Tables 6.1 and 6.2. Overall, the process can be described as a fairly typical A^2O process. While the total SRT is 12 days, the aerobic SRT of the mixed liquor is only 7.2 days. For a temperature of 10°C, this is well above the minimum SRT required to achieve an effluent ammonia below 1 mg N/L, as per the CAS design curves presented in Chapter 1, see Figure 1.1. As per Figure 1.2, however, the ratio of maximum potential nitrification activity to average influent load, $R_{Nit,max/L}$, is approximately 1.5. We would therefore expect to see effluent breakthrough of ammonia under peak loads approximately equal to or greater than 1.5 times the average load. As will be demonstrated in the next section, this certainly is the case.

Table 6.1
Flows, Loads and Plant Operating Conditions for Investigation of Operational Dynamics

Item	Units	Value
Average of Influent Patterns		
Flow	m^3/d	7,902
TKN	mg N/L	32
Ammonia	mg N/L	19
COD	mg/L	392
TSS	mg/L	181
Influent Peaking Factors		
Hourly Flow	–	1.57
Daily Flow	–	1.25
Hourly Load	–	3.64
Daily Load	–	1.46
Plant Operating Conditions (constant)		
RAS Pacing	m^3/d	80%
Total SRT	d	12
Aerobic SRT	d	7.2
Temperature	°C	10
[1]Ammonia Removed in Biofilm, $F_{Nit,B}$	–	30%

[1]Only applies to the MABR/AS process configuration.

Table 6.2
Bioreactor Configuration for Investigation of Operational Dynamics

Item	Units	Tank 1	Tank 2	Tank 3	Tank 4	Tank 5
Bioreactor Volume	m^3	436	436	452	452	452
Nominal HRT, V/Q_{Inf}	hours	1.3	1.3	1.4	1.4	1.4
HRT, $V/(Q_{Inf}+Q_{RAS})$	hours	0.7	0.7	0.8	0.8	0.8
Dissolved Oxygen	mg/L	0	0	2	2	2

6.3.2 DYNAMIC PERFORMANCE OF FULL-SCALE SYSTEM

Dynamic results for the process configurations described in Figure 6.9, Tables 6.1 and 6.2, and under the loading patterns presented in Figure 6.10 were run to evaluate the benefit of the following two factors:

1. Presence of media (MABR/AS) *vs.* no MABR media (CAS) in Tank 2, and
2. Oxygen-limited *vs.* ammonia-limited kinetics in the biofilm.

Effluent Ammonia Breakthroughs

Selected results are presented in Figure 6.11 for a seven-day period when the diurnal peaking factors were particularly high. Instantaneous peaks of ammonia in the effluent reached 14 mg N/L in the effluent of the CAS process on Day 31 *vs.* 13.5 and 11 mg N/L for the oxygen-limited and ammonia-limited MABR/AS processes, respectively. The average effluent ammonia concentrations for Day 31 are more moderate,

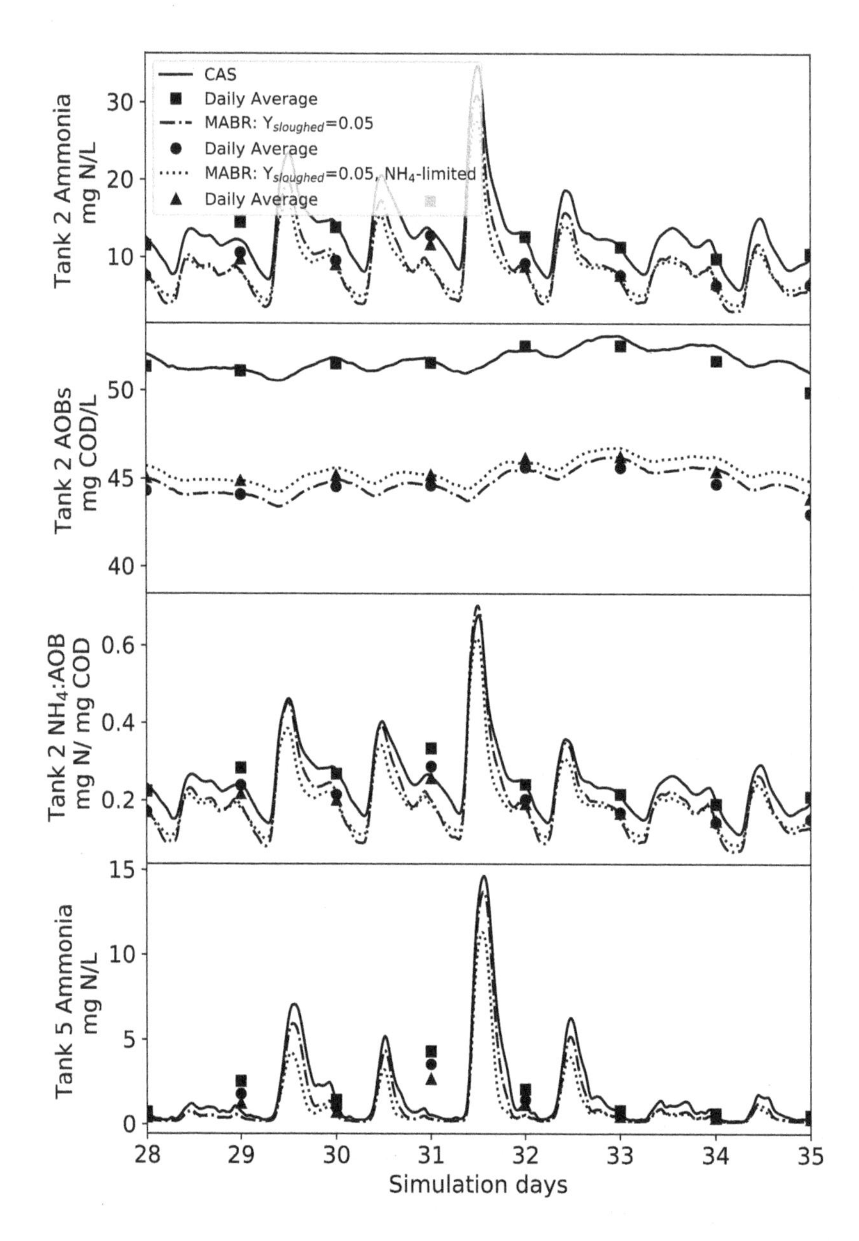

Figure 6.11 Timeseries of secondary effluent ammonia for CAS as compared to MABR/AS under oxygen-limited and ammonia-limited conditions at a temperature of 10°C, SRT = 12 d and $Y_{sloughed}$ = 0.05

but still high, at 4.5, 4 and 3 mg N/L, respectively, for the CAS, oxygen-limited MABR/AS and ammonia-limited MABR/AS.

Observing the benefit provided by the MABR/AS process to reduce effluent ammonia peaks at a relatively high SRT of 12 days (7.2 days aerobic SRT) highlights the importance of evaluating processes under realistic dynamic conditions. Recall from Chapter 4 that, under steady-state conditions, the predicted improvement in effluent quality from removing a fraction of the ammonia load in the biofilm is progressively lower for mixed liquor SRTs above washout. From Figure 4.7 we saw that the difference in predicted effluent ammonia concentration for the hybrid process at $F_{Nit,B}$ = 30%, compared to the conventional alternative ($F_{Nit,B}$ = 0%), is less than 1 mg N/L. In comparison, the dynamic simulations show that on the peak loading day, the MABR/AS process provides peak shaving of 1.5 mg N/L for the day, and 3 mg N/L on an instantaneous basis.

Load Dampening

To better understand the load dampening effects of the hybrid process, Figure 6.11 includes plots of the Tank 2 ammonia concentrations, Tank 2 mixed liquor nitrifying organism concentration (AOBs), and the ratio of the two (NH_4:AOB). Since the effluent from Tank 2 is the influent to Tank 3, and Tank 3 is the first aerobic zone where mixed liquor nitrification can take place, these plots provide insight into the impact of the biofilm on mixed liquor nitrification.

For the CAS scenario, where there is no biofilm, the Tank 2 ammonia is as would be expected following two unaerated tanks where mixed liquor nitrification is fully inhibited by lack of dissolved oxygen. It reflects the influent ammonia concentration pattern in Figure 6.10 plus the effect of dilution from the return activated sludge. In contrast, the MABR/AS scenarios show a Tank 2 ammonia concentration that is proportionally lower based on the fraction of ammonia removed in the biofilm, $F_{Nit,B}$=30%. This works out to an ammonia removal of about 3 to 5 mg N/L. The work that the mixed liquor in aerobic Tanks 3, 4 and 5 needs to do to is therefore proportionally decreased by 3 to 5 mg N/L.

Although the mixed liquor needs to do less "work" in the MABR/AS scenarios, there are also fewer nitrifying organisms present to do the work. This is shown in the plot of Tank 2 AOBs in Figure 6.11 and is a natural consequence of ammonia removal in the biofilm decreasing the amount of ammonia feeding growth of nitrifiers in the mixed liquor. So then what is the advantage of the hybrid process? The advantage is that the ratio of ammonia concentration to nitrifiers (Tank 2 NH_4:AOB) is lower in the MABR/AS. It is true that there are fewer nitrifying organisms in the mixed liquor of the MABR/AS scenario, but the ammonia load that the mixed liquor needs to remove is also proportionally less.

Benefits of the Ammonia-limited Biofilm

Understanding that a lower Tank 2 NH_4:AOB ratio in an MABR/AS process means that the mixed liquor needs to do proportionally less work, the peak trimming observed in the Tank 5 ammonia concentration makes sense. And the improved peak trimming of the ammonia-limited, as compared to the oxygen-limited, biofilm also can be explained. Both the oxygen-limited and ammonia-limited biofilms remove approximately the same amount of ammonia under average loading conditions. This is clear from comparing the mixed liquor Tank 2 AOBs for the two scenarios, as compared to the CAS. The advantage of ammonia-limited biofilm really becomes apparent during the peak loading events when the ammonia-limited is able to "rise to the occasion" of the higher loading rates.

6.4 AN ALTERNATIVE WAY TO ACHIEVE "SAFETY FACTOR"

6.4.1 QUANTIFYING SAFETY FACTOR IN HYBRID SYSTEMS

The ratios of Tank 2 NH_4:AOB presented in Figure 6.11 are convenient indicators of the load dampening effects of hybrid MABR/AS processes. They are easy to develop based on model simulation outputs of ammonia and AOB concentrations in Tank 2. Because lower ratios contribute to less effluent ammonia breakthrough in Tank 5, this ratio can be seen as another proxy for safety factor, where a lower Tank 2 NH_4:AOB ratio is better. In constrast, safety factor was discussed in Chapters 1 and 4 in terms of the steady-state ratio of maximum potential nitrification activity to load in the mixed liquor, $R_{Nit,max/L}$. In this case, a higher ratio of $R_{Nit,max/L}$ is better. In fact Tank 2 NH_4:AOB and $R_{Nit,max/L}$ are two sides of the same coin. $R_{Nit,max/L}$ could be calculated as the reciprocal of Tank 2 NH_4:AOB multiplied by the maximum specific nitrification rate μ/Y. The calculation of $R_{Nit,max/L}$ was presented in Chapter 1, and then again in Chapter 4, and its advantages are described as follows:

- Provides a means to quantify the ratio of mixed liquor nitrifying organisms to ammonia load that accounts for increased endogenous decay at longer SRTs.
- Provides an equivalent means for comparing safety factor between conventional and hybrid biofilm processes.

Note that $R_{Nit,max/L}$, like all the design equations and metrics developed in Chapters 1 and 4, is based on steady-state assumptions. It therefore does not account for the dynamic load dampening benefits of ammonia-limited biofilms. The appropriate tool for quantifying these is dynamic simulation. For the above case study, this means tracking the mixed liquor ammonia load to mixed liquor nitrifier inventory in Tank 2 and, of course, the final effluent ammonia.

While dynamic simulations and $R_{Nit,max/L}$ are useful for understanding the benefits of biofilms during design or in a process optimization study, can they be of any use for operations? Is there a way to monitor the safety factor or ability to manage peak loads in real-time? Figure 6.2 provides some guidance to answering these questions. If ammonia-limited conditions are the key to managing peak loading events, then real-time monitoring of the fraction of the bioreactor that is "ammonia-limited" could provide the means for quantifying safety factor. This could mean using sensors to quantify the fraction of the bioreactor where ammonia concentrations are in a range that is rate limiting, from 0 to approximately 1 or 2 mg N/L for mixed-liquor, and from 0 to as high as 20 mg N/L for MABR biofilms.

Ammonia-based aeration control (ABAC) may provide an even more elegant means for quantifying safety factor. To illustrate, an ABAC control strategy that, under average loading conditions, results in a DO setpoint of 1 mg O_2/L clearly has capacity to spare: turning up the DO setpoint from 1 to 2 mg O_2/L will enhance nitrification capacity during peak loading events. Assuming a Monod relationship to describe the effects of dissolved oxygen on nitrification rate, the safety factor of the process can be quantified based on how low the DO setpoint can drop during average loading conditions. This methodology would still be valid where ABAC was applied in a hybrid system.

6.4.2 LONGER SRTS OR MORE BIOFILMS?

Achieving reliable nitrification in activated sludge has traditionally been about maintaining as long an SRT as possible to maximize the inventory of nitrifiers available to remove a variable incoming ammonia load. Seen from another angle, the goal has been to minimize the load of ammonia relative to the inventory of nitrifying organisms present in the mixed liquor. This has been presented alternately in this book as the mixed liquor NH_4:AOB, Figure 6.11, or as in Chapters 1 and 4, the maximum potential nitrification activity to load in the mixed liquor, $R_{Nit,max/L}$.

Minimizing NH_4:AOB, or maximizing $R_{Nit,max/L}$, by operating at extended SRTs provides added safety factor in managing peak loading events. However, there are limits to this philosophy. There are diminishing returns to operating at longer SRTs because of endogenous decay of nitrifiers in the mixed liquor. Longer SRTs means more decay of the nitrifiers. It is important to recognize how hybrid systems achieve safety factor through the combined impacts of ammonia removal in the biofilm, the seeding effect and, in the case of ammonia-limited biofilms, natural load dampening. Hybrid biofilm processes thus provide an alternative way to achieve nitrification "safety factor" that does not rely on extended SRTs.

As discussed in the opening chapter, the goal of process intensification should not be to reduce safety factor. A safety factor is critical for managing incoming load variations and providing the flexibility required to operate the plant. Operating the activated sludge process at a reduced safety factor is not really process intensification at all, it simply increases the risk of effluent compliance exceedences. But the opportunity cost of achieving safety factor through extended SRTs cannot be ignored. This holds true whether you are considering capital and operating costs, or long-term strategies for resource recovery of the carbon and nutrients in the influent wastewater. By enabling enhancement of safety factor at lower SRTs, hybrid biofilm processes offer a valuable alternative to conventional activated sludge that achieves what would otherwise be incompatible goals of process reliability, low capital and operating costs, and a roadmap to resource recovery.

A Glossary

Ammonia-limited kinetics: Also referred to as substrate-limited kinetics, is said to occur when increasing bulk ammonia loading is accompanied by an increase in ammonia removal rate.

ASM models: Term used by practitioners to refer to a family of biokinetic models initially developed for the activated sludge process. These models are used both in research and commercial simulation software and have since been applied to predict the behavior of biomass in mixed liquor and biofilms.

Biochemical oxygen demand (BOD): Is used in this text to refer to organic material that is biodegradable and will exert an oxygen demand in activated sludge mixed liquor or biofilm. It should not be confused with BOD_5, which refers to the portion of this material that will exert an oxygen demand under laboratory conditions in a five-day test.

Capacity: The quantity of flow, load or population equivalent that a process can treat.

Completely-mixed reactor: Refers to a reactor in which influent is continuous and effluent flow exactly matches influent. The concentrations of soluble and particulate components in the effluent to the reactor are exactly equal to their concentrations inside the reactor which is completely homogenous. The CSTR is the fundamental assumption of all process modeling except for batch processes where both feed and effluent flow are zero.

Conventional Activated Sludge (CAS): Refers to the traditional activated sludge process in which the bioreactor does not contain any media for biofilm growth and unaerated zones have not been created for biological nutrient removal.

Fixed-media/AS: An IFAS configuration in which the biofilm component is provided by a fixed-media supported biofilm.

Flux: Flux is a term used to described the rate of soluble components diffusing into and out of a biofilm. The ammonia flux into a biofilm may be used synonymously with biofilm nitrification rate.

Fully-penetrated: A biofilm is said to be fully-penetrated with substrate or oxygen when there are non-limiting concentrations from the outside to the inside of the biofilm.

Integrated Fixed-Film/Activated Sludge (IFAS): Inclusion of media supported biofilm within the bioreactor tanks of the activated sludge process. In this process, sludge in the biofilm provides complementary treatment to the sludge in the mixed liquor.

Intensification: The term process intensification refers to increasing the quantity of flow or load that can be treated within an existing volume.

Membrane Aerated Biofilm Reactor (MABR): A reactor in which a media supported biofilm is grown with oxygen diffusing through the media into the base of the biofilm. The result is a counter-diffusional biofilm with substrate diffusing into the biofilm from the bulk liquid and oxygen diffusing into the biofilm at its base.

Membrane Aerated Biofilm Reactor / Activated Sludge (MABR/AS): An IFAS configuration in which the biofilm component is provided for in an MABR.

Monod relationship: Mathematical expression linking ammonia and/or dissolved oxygen (DO) concentration, and their respective half saturation concentrations, to the nitrification rate: $\mu = \hat{\mu}\frac{DO}{DO+K_{O_2}}\frac{S_{NHx}}{S_{NHx}+K_N}$.

Moving Bed Biofilm Reactor (MBBR): A reactor in which media is retained using sieves. The media supports the growth of a biofilm and is considered "mobile" due to mixing from aeration or mechanical mixing.

Moving Bed Biofilm Reactor / Activated Sludge (MBBR/AS): IFAS configuration in which the biofilm component is provided for in an MBBR.

Oxygen-limited kinetics: Is said to occur when increasing bulk oxygen concentration (or MABR airflow) is accompanied by an increase in substrate removal or OTE.

Oxygen transfer efficiency (OTE): The efficiency with which oxygen is transferred into the process under field conditions. For an MABR this is calculated based on the percentage of oxygen in the exhaust air under field conditions. For bulk liquid diffused aeration, can be calculated based on percentage of oxygen measured in exhaust hoods placed over the bioreactor. More typically, however, calculated based on standard oxygen transfer efficiency measured by diffuser vendor under clean water conditions. Correction factors are then applied to account for field conditions.

Oxygen transfer rate (OTR): The rate at which oxygen is transferred into the biofilm of an MABR. Calculated based on OTE and airflow and often normalized to surface area in $g/m^2/d$.

Rated capacity: The quantity of flow, load or population equivalent that a process is allowed to treat.

Safety factor: The safety factor is multiplied by the minimum SRT to give the design SRT, which serves as the basis for sizing bioreactors and clarifiers. A safety factor is critical for management of influent load variations and to provide the flexibility required to operate the activated sludge process under real-world conditions.

Seeding effect: Describes the beneficial impact to the mixed liquor nitrification capacity from nitrifying organisms detached or sloughed from a biofilm. May be used interchangeably with "bioaugmentation".

Sludge Retention Time (SRT): V/Q. Note that aerobic SRT is the critical parameter for nitrification. This is relevant in BNR processes that can maintain significant unaerated portions of the bioreactor.

Trickling Filter/Activated Sludge (TF/AS): A process in which trickling filters are followed by activated sludge process.

Two-step nitrification: A model in which two distinct groups of autotrophic organisms sequentially convert ammonia to nitrite and then nitrate. This is the state-of-the-art approach for modeling nitrification in simulation software.

Washout SRT: Below the washout SRT, nitrifying organisms are leaving the system at a rate faster than they are multiplying. This is unsustainable so we speak of a "washout" condition.

B List of Symbols

A: The surface area of media, m^2

AE: Aeration efficiency expressed as the kg of oxygen delivered to the process per kWh of electricity consumed, kg O_2/kWh.

$A_{Nit,max}$: Maximum potential nitrification activity of the mixed liquor assuming neither ammonia nor oxygen are rate limiting, kg NH_4^+-N/d

b: Specific decay rate of ammonia oxidizing bacteria, d^{-1}

b_{20C}: Specific decay rate of ammonia oxidizing bacteria at 20°C, d^{-1}

B_{Solids}: Dry mass of biofilm coverage on the media, g/m^2

$F_{Nit,B}$: The fraction of the influent ammonia load removed in the biofilm, %

F_{Fill}: The fill fraction of media in a bioreactor tank in an MBBR/AS process, %

J_N: Using the design equations of Sen & Randall presented in Chapter 2, and for given bulk liquid dissolved oxygen and ammonia concentrations, J_N is the maximum potential flux of ammonia into a co-diffusional biofilm, g $N/m^2/d$

$J_{N,max}$: Using the design equations of Sen & Randall presented in Chapter 2, and for a given bulk liquid dissolved oxygen concentration, $J_{N,max}$ is the maximum potential flux of ammonia into a co-diffusional biofilm, g $N/m^2/d$

J_{O_2}: The flux of oxygen into a biofilm, $g/m^2/d$

$k_{n,BF}$: Using the design equations of Sen & Randall presented in Chapter 2, the bulk liquid ammonia concentration at which nitrification rate *in the biofilm* is reduced from its maximum potential by a factor of 2, mg N/L

K_S **or** K_N: The bulk liquid ammonia concentration at which nitrification rate of the mixed liquor is reduced from its maximum potential by a factor of 2, mg N/L

L_{ML}: The load of ammonia that the mixed liquor must treat after removal of a portion in the biofilm, kg NH_4^+-N/d

M_{AOB}: The mass of ammonia oxidizing bacteria (nitrifiers) in the bioreactor mixed liquor, kg COD

$M_{Biofilm}$: The total mass of dry solids in the biofilm, kg

M_{MLSS}: The total mass of mixed liquor suspended solids, kg

$MLSS$: The concentration of mixed liquor suspended solids, mg/L

N: The concentration of ammonia in the bioreactor, mg N/L

Q: Infuent flow, m^3/d

$R_{Nit,max/L}$: The maximum potential nitrification activity of the mixed liquor as a ratio to the average influent load that the mixed liquor must treat, kg NH_4^+-N/d per kg NH_4^+-N/d

S **or** S_{NHx}: The concentration of ammonia in the bioreactor which is equal to the concentration of ammonia in the effluent under completely mixed conditions, mg N/L

SF: The safety factor applied to the design SRT, –

S_0 **or** $S_{NHx,0}$**:** The effective concentration of ammonia that the mixed liquor must treat after removal of a fraction, $F_{Nit,B}$, by the biofilm, mg N/L

S_{Inf} **or** $S_{NHx,Inf}$**:** The influent concentration of ammonia prior to removal of a fraction, $F_{Nit,B}$, by the biofilm, mg N/L

SRT**:** The sludge retention time, d

SRT_B **or** SRT_{BF}**:** The average retention time of solids in the biofilm, d

SRT_{Design}**:** The sludge retention time that serves as the basis for sizing the bioreactor tanks and secondary clarifier and which includes safety factor, d

SRT_{min}**:** The minimum SRT required to achieve an effluent ammonia objective using steady-state design equations and without considering any safety factor, d

SSA**:** The specific surface area of MBBR media, m^2/m^3

T**:** Temperature, °C

V**:** The volume of the bioreactor, m^3

X_{AOB} **or** X**:** The concentration of ammonia oxidizing bacteria (nitrifiers) in the bioreactor, mg COD/L

$X_{AOB,0}$ **or** X_0**:** The equivalent concentration of ammonia oxidizing bacteria (nitrifiers) in the influent when considering seeding from a nitrifying biofilm upstream, or integral to, the activated sludge process, mg COD/L

Y**:** The growth yield of ammonia oxidizing bacteria (nitrifiers), mg COD/mg N

$Y_{obs.}$**:** The observed yield of ammonia oxidizing bacteria (nitrifiers) including effects of endogenous decay, mg COD/mg N

$Y_{sloughed}$**:** The yield of ammonia oxidizing bacteria (nitrifiers) sloughed from the biofilm relative to the ammonia removed in the biofilm, mg COD/ mg N

$\hat{\mu}_{20C}$**:** The maximum specific growth rate of ammonia oxidizing bacteria (nitrifiers) at 20°C when neither ammonia nor oxygen are limiting, d^{-1}

μ**:** Used in this book to represent the specific growth rate of ammonia oxidizing bacteria (nitrifiers) in the mixed liquor with a dissolved oxygen concentration of 2 mg O_2/L and assuming ammonia is non-limiting, d^{-1}

τ**:** Used to denote hydraulic retention time which, in a chemostat reactor, is equivalent to the solids retention time (SRT). This symbol is used as an abbreviation for HRT in derivation of analytical solutions, d

θ_b**:** The Arrhenius coefficient that describes the effect of temperature on ammonia oxidizing bacteria (nitrifiers) specific decay rate b, –

θ_μ**:** The Arrhenius coefficient that describes the effect of temperature on ammonia oxidizing bacteria (nitrifiers) specific growth rate μ, –

References

1. Metcalf & Eddy AECOM. *Wastewater Engineering: Treatment and Resource Recovery, Fifth Edition*. McGraw-Hill Education, 2014.
2. M. Aybar, P. Perez-Calleja, J.P. Pavissich, and R. Nerenberg. Predation creates unique void layer in membrane-aerated biofilms. *Water Research*, 149:232–242, 2019.
3. A.C. Cole, M.J. Semmens, and T.M. Lapara. Stratification of activity and bacterial community structure in biofilms grown on membranes transferring oxygen. *Appl Environ Microbiol*, 70(4):1982–1989, 2004.
4. P. Côté, J.-L. Bersillon, and A. Huyard. Bubble-free aeration using membranes: mass transfer analysis. *Journal of Membrane Science*, pages 91–106, 1989.
5. G.T. Daigger, L.E. Norton, R.S. Watson, D. Crawford, and R.B. Sieger. Process and kinetic analysis of nitrification in coupled trickling filter/activated sludge processes. *Water Environment Research*, 65(6):750–758, 1993.
6. G.T. Daigger, E. Redmond, and L. Downing. Enhanced settling in activated sludge: design and operation considerations. *Water Science & Technology*, 78:247–258, 2018.
7. M.K. de Kreuk and M.C.M. van Loosdrecht. Aerobic granular sludge - state of the art. *Water Science & Technology*, 55(8):75–81, 2007.
8. L.S. Downing and R. Nerenberg. Effect of bulk liquid BOD concentration on activity and microbial community structure of a nitrifying, membrane-aerated biofilm. *Applied Microbiology and Biotechnology*, 81:153–162, 2008.
9. Water Environment Federation. *Biofilm Reactors WEF Manual of Practice No. 35*. Water Environment Federation, Alexandria, VA, USA, 2010.
10. C.P.L. Jr. Grady, G.T. Daigger, N.G. Love, and C.D.M. Filipe. *Biological Wastewater Treatment*. CRC Press, Taylor & Francis Group, 2011.
11. L. Hem, B. Rusten, and H. Ødegaard. Nitrification in a moving bed biofilm reactor. *Water Research*, 28(6):1425–1433, 1994.
12. M. Henze, W. Gujer, T. Mino, and M. van Loosdrecht. *Activated Sludge Models ASM1, ASM2, ASM2d and ASM3*. London: IWA Publishing, 2000.
13. H. Horn and S. Lackner. Modeling of biofilm systems: a review. *Adv. Biochem. Eng. Biotechnol.*, pages 53–76, 2014.
14. D. Houweling, Z. Long, J. Peeters, N. Adams, P. Côté, G. Daigger, and S. Snowling. Ntirifying below the "washout SRT": experimental and modeling results for a hybrid MABR / activated sludge process. In *Proceedings of the WEFTEC 2018*, pages 1250–1263. WEFTEC Press, 2018.
15. D. Houweling, F. Monette, L. Millette, and Y. Comeau. Modelling nitrification of a lagoon effluent in moving-bed biofilm reactors. *Water Quality Research Journal of Canada*, 42(4):284–294, 2007.
16. D. Houweling, J. Peeters, P. Côté, Z. Long, and N. Adams. Proving membrane aerated biofilm reactor (MABR) performance and reliability: results from four pilots and a full-scale plant. In *Proceedings of the WEFTEC 2017*. WEFTEC Press, 2017.
17. C. Jenkins and S.J. Yeh. Pure oxygen fixed film reactor. *Journal of the Environmental Engineering Division, ASCE*, 4:611–623, 1978.
18. J. Jimenez, D. Dursun, P. Dold, J. Bratby, J. Keller, and D. Parker. Simultaneous nitrification-denitrification to meet low effluent nitrogen limits: modeling, performance and reliability. In *Proceedings of the WEFTEC 2010*. WEFTEC Press, 2010.

19. T.E. Kunetz, A. Oskouie, A. Poonsapaya, J. Peeters, N. Adams, Z. Long, and P. Côté. Innovative membrane-aerated biofilm reactor pilot test to achieve low-energy nutrient removal at the Chicago MWRD. In *Proceedings of the WEFTEC 2016*. WEFTEC Press, 2016.
20. T.M. LaPara, A.C. Cole, J.W. Shanahan, and M.J. Semmens. The effects of organic carbon, ammoniacal-nitrogen, and oxygen partial pressure on the stratification of membrane-aerated biofilms. *J Ind Microbiol Biotech*, 33(4):315–323, 2006.
21. Ontario Ministry of the Environment Sewage Technical Working Group. *Design Guidelines for Sewage Works*. Ontario Ministry of the Environment, 2008.
22. Wastewater Committe of the Great Lakes Upper Mississippi River. *Recommended Standards for Wastewater Facilities (Ten States Standards)*. Board of State and Provincial Public Health and Environmental Managers, 2014.
23. C. Pellicer-Nàcher, S. Sun, S. Lackner, A. Terada, F. Schreiber, Q. Zhou, and B.F. Smets. Sequential aeration of membrane-aerated biofilm reactors for high-rate autotrophic nitrogen removal: experimental demonstration. *Environmental Science and Technology*, 44(19):7628–34, 2010.
24. E. Plaza, J. Trela, and B. Hultman. Impact of seeding with nitrifying bacteria on nitrification process efficiency. *Water Science & Technology*, 43:155–163, 2001.
25. German ATV-DVWK Rules and Standards. *Dimensioning of Single-Stage Activated Sludge Plants*. German Association for Water, Wastewater and Waste, 2000.
26. B. Rusten, L. Hem, and H. Ødegaard. Nitrification of municpal wastewater in novel moving bed biofilm reactors. *Water Environment Research*, 67(1):75–86, 1995.
27. K. Rutt, J. Seda, and C.H. Johnson. Two year case study of integrated fixed film activated sludge (ifas) at Broomfield, CO WWTP. In *Proceedings of the WEFTEC 2006*, pages 225–239, 2006.
28. R. B. Schaffer, F. J. Ludzack, and M. B. Ettinger. Sewage treatment by oxygenation through permeable plastic films. *Journal (Water Pollution Control Federation)*, 32(9):939–941, 1960.
29. D. Sen and C.W. Randall. Improved computation model (Aquifas) for activated sludge, integrated fixed-film activated sludge, and moving-bed biofilm reactor systems, part i: semi-empirical model development. *Water Environment Research*, 80(5):439–453, 1960.
30. P.S. Stewart. Diffusion in biofilms. *Journal of Bacteriology*, 185(5):1485–1491, 2003.
31. N. Sunner, Z. Long, D. Houweling, and J. Monti, A. and Peeters. MABR as a low-energy compact solution for nutrient removal upgrades – results from a demonstration in the UK. In *Proceedings of the WEFTEC 2018*, pages 1264–1281. WEFTEC Press, 2018.
32. I. Takács, C.M. Bye, K. Chapman, P.L. Dold, P.M. Fairlamb, and R.M. Jones. A biofilm model for engineering design. *Water Science & Technology*, pages 1–8, 2007.
33. W.A. Thomas. *Evaluation of Nitrification Kinetics for a 2.0 MGD IFAS Process Demonstration*. Master's thesis from Virginia Polytechnic Institute, Blacksburg, VA, USA, 2009.
34. A. Underwood, C. McMains, D. Coutts, J. Peeters, J. Ireland, and D. Houweling. Design and startup of the first full-scale membrane aerated biofilm reactor in the United States. In *Proceedings of the WEFTEC 2018*, pages 1282–1296. WEFTEC Press, 2018.
35. O. Wanner and P. Reichert. Mathematical modeling of mixed-culture biofilms. *Biotechnology & Bioengineering*, 49:172–184, 1996.
36. H. Ødegaard. Applications of the MBBR processes for nutrient removal. In *Proceedings of the 2nd Specialized IWA Conference in Nutrient Management in Wastewater Treatment Processes*, Kraków, Poland, 6-9 September 2009, pages 1–11, 2009.
37. Europe, S. 2019. Surfrider Foundation Europe. Available at: https://surfrider.eu/en/biomedia-filters-a-new-pollution-in-the-atlantic/ Accessed 3 Jul 2019.

Index